Challenges in the Binational Management of Water Resources in the Rio Grande/Río Bravo

by
David J. Eaton
and
David Hurlbut

A report based on a research project of the
U.S.-Mexican Policy Studies Program
Sidney Weintraub, Director
Jan Gilbreath, Series Editor

U.S.-Mexican Policy Report No. 2
Lyndon B. Johnson School of Public Affairs
The University of Texas at Austin
1992

© 1992 by the Board of Regents, The University of Texas System

Published in 1992 in the United States of America by the Lyndon B. Johnson School of Public Affairs, The University of Texas at Austin, P.O. Drawer Y, University Station, Austin TX 78713-7450

Report series editor: Jan Gilbreath
Book design: Doug Marshall, LBJ School Office of Publications

Challenges in the Binational Management of Water Resources in the Rio Grande/Río Bravo
by David J. Eaton and David Hurlbut

ISBN 0-89940-315-8
Library of Congress Catalog Card No. 92–82935

U.S.-Mexican Policy Studies Program

Policy Report Series

Policy Report No. 1: *Free Trade with Mexico: What's in It for Texas?*

Policy Report No. 3: *Policymaking, Politics, and Urban Governance in Chihuahua: The Experience of Recent Panista Governments*

Policy Report No. 4: *Health Care Across the Border: The Experience of U.S. Citizens in Mexico*

Policy Report No. 5: *U.S.-Mexican Environmental Policy: Who's Polluting Whom?*

Policy Report No. 6: *Motor Carrier Regulatory Reform Under the Salinas de Gortari Administration*

Occasional Papers

No. 1: *Planning the Border's Future: An Analysis of the Mexican-U.S. Environmental Plan*

No. 2: *Australia and New Zealand: The Challenge of NAFTA*

Related Publications

Special Project Report: *The U.S.-Mexico Free Trade Agreement: Economic Impact on Texas,* 1992.

Policy Research Project Report No. 98: *Texas-Mexico Transborder Transportation System: Regulatory and Infrastructure Obstacles to Free Trade,* 1991.

Policy Research Project Report No. 93: *Maximizing Benefits of Tourism in Guerrero, Mexico,* 1991.

For order information, call (512) 471-8951 or write to the U.S.-Mexican Policy Studies Program, Lyndon B. Johnson School of Public Affairs, The University of Texas at Austin, P.O. Drawer Y, University Station, Austin, TX 78713-7450.

Contents

Figures

Tables

Contributors

David J. Eaton is Bess Harris Jones Centennial Professor in Natural Resource Policy Studies at the Lyndon B. Johnson School of Public Affairs, The University of Texas at Austin. He received his Ph.D. in Geography and Environmental Engineering from The Johns Hopkins University in 1977. He has done extensive research and consulting on water resource management in Texas and in other regions of the United States, as well as in China, the Middle East, Latin America and Eastern Europe.

David Hurlbut is a policy analyst and doctoral student at the Lyndon B. Johnson School of Public Affairs. His area of emphasis is international development and environmental policy. He received his master's in public affairs at the LBJ School in 1991. Prior to that, he was a staff reporter for the *Dallas Times Herald*. He is co-author of another monograph in this series, *Free Trade with Mexico: What's in It for Texas?*

Preface

THE PURPOSE OF THIS MONOGRAPH IS TO EXAMINE emerging water conflicts along the Texas-Mexico border, describe current efforts to resolve them, and identify possible solutions. This monograph is optimistic in tone, perceiving realistic alternatives for existing surface water, groundwater, and water quality problems. But the solutions require money, significant institutional change, and modifications of behavior of people in the region. Chapter One reviews the current patterns of water use along the Rio Grande/Río Bravo by nation and use sector. Chapter Two examines the pattern of surface water withdrawals over time in the context of existing Mexico-U.S. treaty obligations. Chapter Three addresses groundwater resources and the problem of over-extraction of these underground commons. Chapter Four is concerned with surface water quality and the consequences of untreated sewage on border health and the environment. Chapter Five investigates alternative institutional arrangements which facilitate the resolution of these regional problems. Chapter Six presents results of a significant survey of public opinion regarding water issues in one of the twin cities along the border, Nuevo Laredo/Laredo.

The research for this monograph was undertaken under the auspices of the U.S.-Mexican Policy Studies Program at the Lyndon B. Johnson School of Public Affairs, University of Texas at Austin. The program incorporates a wide range of research into binational policy issues. Among these are research on the U.S.-Mexican trade relationship, including the labor and environmental impacts of a free-trade agreement, opposition governments in Mexico and their impact on the border region and on the binational relationship, and a variety of studies investigating the effects of industrial integration on the U.S.-Mexican border region. The U.S.-Mexican Policy Studies Program is directed by Dr. Sidney Weintraub. Jan Gilbreath is editor of the U.S.-Mexican Policy Reports. The program and the reports are partly funded through The William and Flora Hewlett Foundation of Menlo Park, California. We gratefully acknowledge the support of The Hewlett Foundation in initiating this publication series.

Acknowledgments

PRELIMINARY RESEARCH FOR THIS MONOGRAPH WAS conducted by seven graduate students, one post-doctoral fellow, and two faculty members in the Lyndon B. Johnson School of Public Affairs of The University of Texas at Austin during the 1989-1990 academic year. The graduate students were Clara Castro, Jaime E. Lizarraga, Lynne McGuire, Joseph Roth, Andrew Sherrill, Eric Stockton, and Zichuan Ye. Professor Mahesh Chaturvedi, Professor David Eaton, and Dr. Ziming Yang were the two faculty and post-doctoral fellow, respectively. A research associate, David Hurlbut, and Professor Eaton undertook significant additional research during 1991 and 1992 to complete this monograph. Chapter Six reflects the draft developed by Lynne McGuire and Clara Castro. As the text reflects primarily the ideas and writing of David Eaton and David Hurlbut, we are listed by name as authors.

The Lyndon B. Johnson School provided in-kind services and the Lyndon Baines Johnson Foundation contributed funds to support this research. Professor Eaton was able to discuss a number of these issues at a conference at Gasparilla Island, Florida, sponsored by The Ford Foundation through a grant to Professor Al Utton of the International Transboundary Resources Center at the School of Law of the University of New Mexico. Manuscript preparation and graphics development were prepared on equipment contributed by the Urban Services Laboratory at the LBJ School of Public Affairs. Final book design and computer desktop publishing were handled by Doug Marshall, LBJ School of Public Affairs, Office of Publications.

The Policy Research Project members express appreciation to

the following individuals for their contributions to this research: Dan Beckett, Allen O. Beinke, Jr., Joaquim Bustamante, David Buzan, Tamara Crail, Hoss Evans, Dianne Mendoza Freeman, Joseph P. Friedkin, Jan Gilbreath, Alberto Gonzalez Jr., José de Jesús Herrera, Helen Ingram, Cruz Ito, Norman D. Johns, Sandra Johnson, Stephen Manning, J.R. Mathis, James E. Moore, Stephen P. Mumme, Anton Papich, Tomás Rodriguez, Fernando Roman, Norma Steffa, Alberto Székely, Albert E. Utton, and Ing. Max Rendon Villarreal. In addition we thank the following people who assisted in the preparation of the manuscript: Gail Bunce, Judy Caskey, Marilyn Duncan, María de la Luz Martínez, Jan Gilbreath, Anne Rohe, and Sidney Weintraub.

Although a substantial number of people have contributed to this effort, any remaining errors or omissions are ours.

David Eaton
David Hurlbut

Chapter 1:
An Overview of Water Conflicts and the River Basin

ISCUSSION OVER THE NORTH AMERICAN FREE TRADE Agreement (NAFTA) has directed Mexican and American policymakers' attention to the border regions. Tariff reductions and increased economic integration will have great impact in these areas. Economic activity has a special resonance for the 2,012-kilometer border between Texas and Mexico, where the shared Rio Grande/Río Bravo is home to an estimated four million people. The region is so poor that four of the ten highest rates of poverty are found in counties along the Texas side of the border; their Mexican neighbors on the other side of the river are even worse off. Although trade and production promise a better life, the economic activity will draw more people to the region, increasing pressures on a fragile semi-arid environment with sparse water supplies. Indeed, the availability and quality of water is likely to affect the implementation of NAFTA, as it is a source of current and potential conflict between Mexico and the United States. The simmering disputes relate to surface water allocation, groundwater extraction, and sewage disposal. Each issue has an origin in prior Mexico-U.S. agreements.

The United States and Mexico have had nearly nine decades of successful cooperation on surface water allocation issues under a 1906 border treaty and a 1944 water treaty. The 1944 treaty allocated all the flow of the Rio Grande and its tributaries between Mexico and the United States. As a result of legislative and administrative action within Texas, all the water flowing along the Texas side of the border has been allocated to individual users within each country. Some historical evidence casts doubts as to whether the United States has lived up to the spirit if not the letter of its treaty commitments to provide adequate water to Mexico. Conflicts between the rights and responsibilities of users pose problems for the future.

Sewage treatment conflicts arise from the unequal economic conditions of Texas and Mexico. The quality of surface water within the Rio Grande/Río Bravo violates the standards of both nations. The cause of pollution is the discharge of sewage and other liquid wastes from the six largest twin cities along the border. Through a combination of federal, state, and local revenue sources, the Texas side has made progress in sewage collection and treatment. A lack of resources in Mexico has undermined efforts to treat all raw sewage prior to disposal into the Rio Grande/Río Bravo.

Competitive circumstances induce Mexicans and Texans to overpump groundwater in the absence of controls in either country. While groundwater is a national resource in Mexico and a commodity owned by landowners in Texas, in neither venue is there incentive to conserve groundwater and to assure a sustainable yield for high-valued uses in the future.

Water issues are basic to the life of border residents, who have testified repeatedly to the U.S. and Mexican governments that they are unhappy with current efforts to solve water problems. These concerns cross all social and economic divisions. Solutions to these problems require money, institutional flexibility, and cooperation among nations, states, and local authorities. A variety of institutional mechanisms are available for resolving these conflicts. The ultimate choice will reflect who designs, implements, and pays for the needed hydraulic infrastructure.

Water Basin Overview

The Rio Grande/Río Bravo originates in the San Juan Mountains of southern Colorado. Flowing 989 kilometers from its headwaters and through the state of New Mexico, it enters Texas about 32 kilometers northwest of El Paso and then continues 2,053-kilometers to the Gulf of Mexico.[1] Past the twin cities of El Paso and Ciudad Juárez, the river forms the boundary between Texas and Mexican states of Chihuahua, Coahuila, Nuevo Leon, and Tamaulipas.

Even though the headwaters are in the United States, almost half of this predominantly arid and semi-arid drainage basin lies in Mexico. This excludes closed drainage areas inside the basin

Figure 1.1
Rio Grande/Río Bravo Diversions and Tributary Inflows
(annual averages in thousand acre-feet)

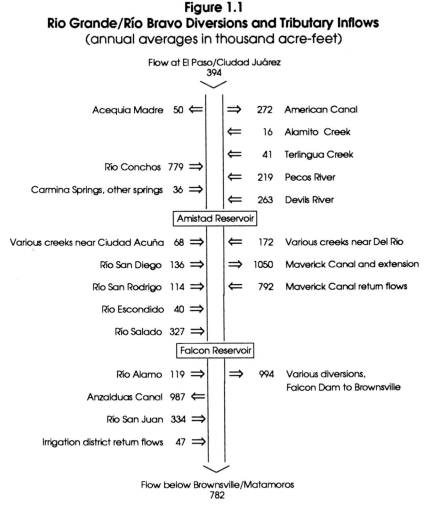

Flow at El Paso/Ciudad Juárez
394

Acequia Madre 50 ⇐	⇒ 272	American Canal
	⇐ 16	Alamito Creek
	⇐ 41	Terlingua Creek
Río Conchos 779 ⇒	⇐ 219	Pecos River
Carmina Springs, other springs 36 ⇒	⇐ 263	Devils River

Amistad Reservoir

Various creeks near Ciudad Acuña 68 ⇒	⇐ 172	Various creeks near Del Rio
Río San Diego 136 ⇒	⇒ 1050	Maverick Canal and extension
Río San Rodrigo 114 ⇒	⇐ 792	Maverick Canal return flows
Río Escondido 40 ⇒		
Río Salado 327 ⇒		

Falcon Reservoir

Río Alamo 119 ⇒	⇒ 994	Various diversions, Falcon Dam to Brownsville
Anzalduas Canal 987 ⇐		
Río San Juan 334 ⇒		
Irrigation district return flows 47 ⇒		

Flow below Brownsville/Matamoros
782

Mexico		United States	
Inflows	2000	Inflows	1503
Diversions	*1037*	*Diversions*	*2316*
Net for Mexico	963	Net for U.S.	- 813

Figures are averages of all years for which data are available at each measurement point; not all averages cover the same time period. Evaporation, groundwater seepage, rainfall, and various minor diversions and inflows are not included. Net of inflows and outflows therefore differs from average flow below Brownsville.

Source: International Boundary and Water Commission, "Flow of the Rio Grande and Related Data," Water Bulletin No. 60 (1990).

that contribute no runoff to the watershed; these closed basins constitute about 47 percent of the nearly 862,500 square kilometers within the watershed's perimeter.[2] Mexico irrigates about 1.1 million acres in the basin, while the United States irrigates about 993,000 acres. Only 98,000 acres of irrigated land lie upstream from Texas. The Conchos, San Rodrigo, Alamo, and San Juan Rivers are the primary tributaries in Mexico. The Pecos and Devils Rivers are the principal tributaries to the river in Texas.

Figure 1.1 depicts the major flows into and withdrawals from the Rio Grande/Río Bravo; Figure 1.2 shows the location of the river basin, major cities, and major reservoirs. Including the inflow at El Paso, an average of about 2.2 billion cubic meters of water flow into the river from the United States, slightly less than the 2.5 billion cubic feet that comes from Mexican sources. The United States generally takes out more water than it puts in.

Much of the main river's flow that reaches El Paso/Ciudad Juárez is removed by irrigation diversions at the American Canal and the Acequia Madre Canal. Most of the river just below the cities consists of treated municipal wastewater and irrigation return flows.[3] About 306 kilometers downstream, the river merges with the Rio Conchos near Presidio. The new inflow regenerates the main trunk of the river, which then flows through the deep canyons of the Big Bend area. The river runs northeast and east through 217 kilometers of limestone canyons and mountains to Langtry, Texas. After passing through low hills and narrow valleys, the river flows into the Amistad Reservoir (completed in 1968) with its total controlled capacity of 6.5 billion cubic meters.[4] The Pecos and Devils Rivers also discharge into the reservoir; the water of the Devils River is of excellent quality, while the Pecos is a major source of salt in the Rio Grande/Río Bravo below the reservoir.[5]

The San Rodrigo River joins the Rio Grande/Río Bravo between Del Rio and Eagle Pass, Texas. Then 128 kilometers below Laredo, Texas, the river flows into Falcon Reservoir (completed in 1953) with a capacity of 4.9 billion cubic meters.[6] The reservoir receives most of its water from the Salado River and is a source of good quality water for the municipalities, farms, and industries of the Lower Rio Grande Valley.[7]

The Río Alamo meets the Rio Grande/Río Bravo about 19 kilometers downstream from Falcon Dam, and the San Juan River enters it about 58 kilometers downstream. Finally, 37 kilome-

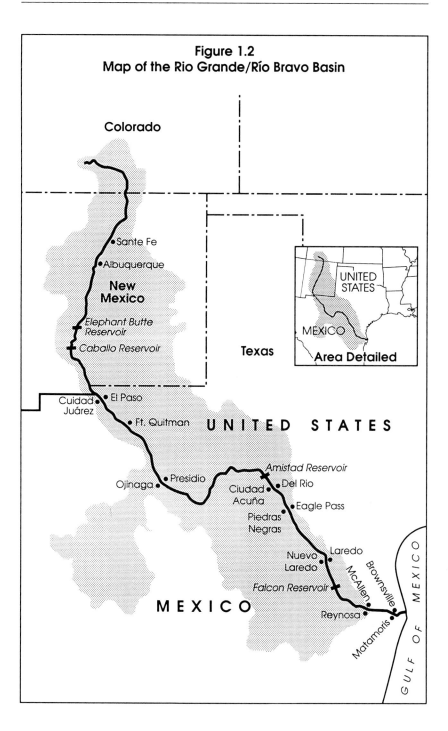

Figure 1.2
Map of the Rio Grande/Río Bravo Basin

ters east of Brownsville and Matamoros, the river empties into the Gulf of Mexico.

The flow of the river varies widely throughout the basin. The average annual discharge 40 kilometers downstream from El Paso/Ciudad Juárez is around 69 million cubic meters, while the average discharge below the Falcon Dam is more than 2.5 billion acre-feet.[8] The upper reach of the river experiences a consistently lower flow rate than either of the other two reaches. However, the middle and lower reaches of the river are subject to substantially greater yearly variations in flow than the upper reach.

Winter temperatures vary considerably from northwest to southeast along the river, although the mean high temperatures for July are between 34.4 and 37.8 degrees Celcius throughout the basin. The warmest areas are near Laredo/Nuevo Laredo and Presidio/Ojinaga. El Paso/Ciudad Juárez has the coldest winters with a mean January low of 0 degree Celcius, while the mean low temperature at the mouth of the river is 10.6 degrees Celcius.[9]

At every section of the Rio Grande, water loss due to evaporation significantly exceeds the amount of water gained through precipitation. Average evaporation is 228 centimeters per year in Ciudad Juárez, while rainfall in the area averages slightly more than 25 centimeters per year. Average annual rainfall is roughly 51 centimeters from Amistad Reservoir to the Gulf of Mexico, but at all measuring points the evaporation rate is still four to five times the amount of rainfall.[10] Using the climatic zone definitions of Meigs, the Rio Grande/Río Bravo basin ranges from arid areas not suitable for crops to semi-arid regions suitable for some crops only, including natural grasslands.[11]

The Texas section of the Rio Grande/Río Bravo River Basin is characterized by a variety of land forms which may be aggregated into three distinctive geographic regions. The upper reach is a region of mountains and basins in which land elevation averages from 305 to 1,524 meters above sea level. The middle reach, which is part of the Edwards Plateau geographic region, is characterized by high hills and average land elevations ranging from 153 meters to 305 meters. The lower reach, part of the South Texas Plain region, contains irregular plains in its northern section and flat plains in its southern section. Average land elevation ranges from 31 meters to 92 meters.

Table 1.1
Population of Metropolitan Areas along the U.S.-Mexican Border

City	1980 population	1990 population	Annual growth rate
Ciudad Juárez, Chi.	567,365	797,679	3.5%
El Paso, Tex.	479,899	591,610	2.1%
Ojinaga, Chi.	26,421	23,947	-1.0%
Presidio, Tex.	1,723	6,637	14.4%
Ciudad Acuña, Chi.	41,948	56,750	3.1%
Del Rio, Tex	30,034	138,721	16.5%
Piedras Negras, Chi.	80,290	98,177	2.0%
Eagle Pass, Tex.	21,407	36,378	5.4%
Nuevo Laredo, Tam.	203,286	219,468	0.8%
Laredo, Tex.	99,285	133,239	3.0%
Reynosa, Tam.	294,934	376,676	2.5%
McAllen, Tex.	283,229	383,545	3.1%
Matamoros, Tam.	238,840	303,392	2.4%
Brownsville, Tex.	209,727	260,120	2.2%

Source: U.S. Census Bureau, Mexican Census Bureau.

On the Texas side of the upper reach of the basin, agricultural land is used primarily for raising cattle and sheep, as well as for growing cotton. Large ranches that raise cattle, sheep, and goats are typical in the middle reach of the basin. Much of the agricultural land in the lower reach is also devoted to raising livestock, as well as cotton and vegetables. At the southern tip of the basin, land is used primarily for growing cotton, vegetables, and citrus fruits.

Major agricultural land uses in the Mexican section of the basin include range livestock, wheat, cotton, and mixed cereals. Land in the upper section of the basin is devoted primarily to raising cattle and sheep. In the lower section, raising livestock is also a major activity, but land is used to grow wheat, cotton, and mixed cereals as well.

The basin's population is highly concentrated in its urban areas. About 1.4 million people live in the El Paso/Ciudad Juárez area—the most populous along the border—and about 57 percent of them are on the Mexican side of the river. Not only does Ciudad Juárez have a larger population than El Paso, it is growing faster. (See Table 1.1.) In the middle reach, Nuevo Laredo is much larger than Laredo (219,000 as compared with 133,000), but the American city is growing at a much faster rate. A high rate of growth has also been seen in the Texas city of Del Rio, which is close to Amistad Reservoir. In the lower reach of the river, McAllen and Reynosa both have about 380,000 people, although McAllen is growing slightly faster.

Notes

1. Raymond Johnson, *The Rio Grande,* East Sussex: Publishers, Inc., 1981, p. 49.

2. International Boundary and Water Commission, "Flow of the Rio Grande and Related Data," Water Bulletin No. 59 (1989), pp. 149-150.

3. Texas Water Development Board, *Water for Texas, Today and Tomorrow,* December 1990, p. III-23-1.

4. Texas Water Commission, "Reconnaissance Investigation of the Ground-Water Resources of the Rio Grande Basin, Texas," Bulletin 6502, Austin, Texas, July 1965, p. U15.

5. TWDB,*Water for Texas,* p. III-23-1.

6. International Boundary and Water Commission, *Joint Projects of the United States and Mexico,* El Paso, Texas, 1981.

7. *Water for Texas,* p. III-23-3; David J. Eaton and John Michael Andersen, *The State of the Rio Grande/Río Bravo,* University of Arizona Press, Tucson, 1987, p. 5.

8. IBWC, "Flow of the Rio Grande and Related Data," Water Bulletin No. 59, 1989.

9. Office of the State Climatologist, as printed in the *Texas Almanac,* The Dallas Morning News, 1987.

10. IBWC, "Flow of the Rio Grande and Related Data," Water Bulletin No. 58, 1988.

11. United Nations Educational, Social and Cultural Organization, *Arid Zone Resource Series No. 1, Arid Zone Hydrology,* "World Distribution of Arid and Semi-Arid Homoclimates," by P. Meigs, 1953.

Chapter 2:
Water Use and Apportionment

MEXICO AND THE UNITED STATES HAVE TWO TREATIES and various cooperative regulations that govern allocation of the water resources they share. In addition, Texas and the upper riparian states of Colorado and New Mexico share a compact governing the often-contentious interstate allocation of the upper basin's scarce water. Within the state of Texas, the Special Water Master serves as a guarantor of equitable allocation among competing local interests. As complex as it may seem, this multijurisdictional framework only mirrors the complexity of the water use issues these institutions were created to address.

At the turn of the century, acute water shortages on the Mexican side of the border in Ciudad Juárez brought water apportionment issues directly onto the bilateral agenda for the first time. The United States had planned to build a dam in New Mexico to impound waters for flood control and irrigation. Mexico, concerned about the reduction of water reaching Ciudad Juárez, sought its rights as a lower riparian to a continual flow of the river. The two nations compromised in the 1906 Convention for the Equitable Division of the Waters of the Rio Grande for Irrigation Purposes, in which Mexico agreed to the development of Elephant Butte Dam in New Mexico and the United States committed to providing Mexico with 74 million cubic meters of river water downstream for Ciudad Juárez and the surrounding area south of the border.

In 1944, the United States and Mexico signed a new treaty establishing cooperative regulation and apportionment of the Rio Grande/Río Bravo from Fort Quitman to the Gulf of Mexico. The pact established two international reservoirs—Amistad and Falcon—and set explicit guidelines governing diversions from the Rio Grande/Río Bravo and its tributaries. This treaty also replaced International Boundary Commission with the International Boundary and Water Commission (IBWC) as the authoritative monitoring and management agency.

Despite the tremendous economic growth that has taken place in the region since they were signed, the two operative international agreements governing water allocation in the Rio Grande/Río Bravo basin continue to be the 1906 Convention and the 1944 Treaty. At the time the 1906 Convention was signed—when El Paso County comprised fewer than 53,000 people—the United States could withdraw all the water it needed and still leave enough in the river to provide for Mexico's entitlement.[1] But today, more than 640,000 people live in El Paso County, and agriculture has increased significantly. Water consumption has increased so much on the U.S. side that the flow of the river from El Paso/Ciudad Juárez to Presidio/Ojinaga dwindles to a trickle—and at times ceases entirely. This has made both El Paso and Juárez more dependent on underground aquifers, but neither the 1906 Convention nor the 1944 Treaty address groundwater apportionment at all.

This chapter discusses water consumption patterns along the Rio Grande/Río Bravo. Among the water allocation issues are the volume of water used on each side of the border, the uses of water withdrawals, and the likelihood of future shortages. Because of the link between water and growth, trends in overconsumption and undersupply indicate problems that could arise if the United States and Mexico continue to integrate their economies without taking steps toward cooperatively managing their shared water resources.

Apportionment and Water Use Patterns

The Texas Water Development Board (TWDB) projects that by the year 2040, the Rio Grande/Río Bravo and its associated tributaries and aquifers will fall about 338 million cubic meters per year short of being able to meet all the demands placed on it by water users on the U.S. side of the river. The bulk of the shortage—211 million cubic meters—is expected to fall on agricultural users.[2]

The United States typically takes twice as much water from the Rio Grande/Río Bravo as Mexico does. On average, Mexico withdraws 1.2 billion cubic meters of water from the river every year—64 million through the Acequia Madre Canal in Ciudad Juárez, and the rest through the Anzalduas Canal near Reynosa. The United States diverts an average of 340 million cubic meters at El Paso, 1.1 million cubic meters in Maverick County, and a total of 1.2 million cubic meters at various points along the lower river below Falcon Dam.[3]

In Texas, municipal needs have priority over agriculture and industry in water apportionment between sectors. Thus, as population grows in cities along the border, growers will find themselves with less water for irrigation once excess water is taken by municipalities. The major cities on the U.S. side of the river (El Paso, Laredo, McAllen-Edinburgh, and Brownsville) are growing at annual rates of between 2 percent and 3 percent. With conservation measures, municipal demand on the U.S. side of the basin is expected to increase 1.3 percent per year during the first half of the next century, accounting for 18 percent of total water demand by 2040.[4] On the Mexican side, Reynosa and Matamoros are growing by about 2.5 percent annually, while Ciudad Juárez is growing by 3.5 percent. In addition, the average Mexican will tend to use more water as incomes rise in these Mexican cities, thus accelerating the rate of overall water consumption.

Yet as projected demand increases, the supply of water available to meet it is expected to diminish. Texas water officials project that for the basin as a whole, aquifer supplies will fall 0.8 percent annually from 2000 to 2040. This will cause particular hardships on El Paso/Ciudad Juárez and other municipalities of the upper basin. Water conservation could help make up for much of the shortfall, but not enough to offset the magnitude of the gap between increasing demand and diminishing supply.

The four largest metropolitan areas in the Rio Grande/Río Bravo Basin will present particular problems for future water planning: El Paso/Ciudad Juárez in the upper basin, Laredo/Nuevo Laredo in the river's middle reach, and McAllen/Reynosa and Brownsville/Matamoros in the lower reach. Population pressures are projected to create water shortages in every area by the year 2040. But for the most part, strong conservation measures could keep demand within available supply.[5] The exception will be El Paso/Ciudad Juárez, where shortages are likely even with the best conservation efforts.

The Upper Reach: El Paso/Ciudad Juárez

Agriculture on both sides of the border contends for nearly all the surface water of the Rio Grande/Río Bravo's upper reach. The 1906 Convention requires the United States to "deliver to Mexico a total of 74 million cubic meters of water annually" except in years of extraordinary drought, and the agreement breaks this annual number down into monthly volumes to reflect seasonal fluctuations in the flow level. As the agreement is worded, whether Mexico actually captures 74 million cubic meters is not a consideration; the United States is only obliged to leave the proscribed amount of water in the river once it has diverted its share.[6]

The volumes specified in the treaty are so small that fulfilling them posed no apparent problems for the first 30 years the Convention was in force. The International Water Commission (later merged into the International Boundary and Water Commission by the 1944 Treaty) monitored the river's flow just above Mexico's Acequia Madre canal at Ciudad Juárez, and then 31 kilometers downstream near Tornillo. These records show that from 1924 to 1937, nearly one-third of the river's average flow of 709.3 million cubic meters was left over after all U.S. and Mexican consumption between the two measuring points.[7] This remained consistent throughout the period for which records are available. When the U.S. dam and canal were opened in 1938, water users on the U.S. side suddenly were able to take 80 percent of the river at once. Since then, ensuring fulfillment of the 1906 Convention has required careful monitoring; the United States has had to invoke the Convention's drought provision 14 times in the past 50 years because it could not provide the requisite 74 million cubic meters of water.

A statistical analysis of water flow data since 1939 (the earliest year for which complete and reliable records are available) shows that apportionment between the United States and Mexico *as a percentage of total flow* has been remarkably consistent. Despite wide fluctuations in the river's initial inflow from year to year, the United States takes about 79 percent of the water that reaches the twin cities. Mexico takes most of what is left over—about 14.4 percent of the initial flow. During a typical year, only 6.6 percent of the river remains by the time it leaves El Paso/Ciudad Juárez.[8]

Figure 2.1 shows the historical pattern of water allocation

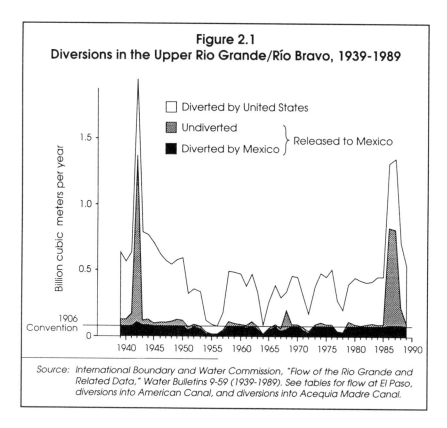

Figure 2.1
Diversions in the Upper Rio Grande/Río Bravo, 1939-1989

☐ Diverted by United States

▨ Undiverted

■ Diverted by Mexico } Released to Mexico

Source: International Boundary and Water Commission, "Flow of the Rio Grande and Related Data," Water Bulletins 9-59 (1939-1989). See tables for flow at El Paso, diversions into American Canal, and diversions into Acequia Madre Canal.

between the United States and Mexico. The U.S. Bureau of Reclamation regulates the amount of water released from Caballo and Elephant Butte Reservoirs in New Mexico, about 100 miles upstream from El Paso/Ciudad Juárez. Water for use on the U.S. side of the border is diverted into the American Canal at El Paso; two miles downstream, Mexico diverts part of the remainder through the Acequia Madre in Ciudad Juárez. Typically, the Mexican canal is empty from mid-September through mid-March as the initial river flow drops to less than 12.3 million cubic meters per month. The United States continues to divert an average of 9.3 million cubic meters per month during this low period. For the rest of the year, Mexico takes an average of 10.7 million of the 19.6 million cubic meters per month remaining in the river after diversion by the United States.[9] The Acequia Madre does have the capacity to carry more water (as much as 21.5 million cubic meters per month), but only if the flow is consistently high. Otherwise Mexico is limited day-to-day and month-to-month by the amount of water diverted by the United States two miles upstream.

Statistical analysis reveals some unusual trends buried in the averages. Even though Mexico generally takes 14.4 percent of the initial flow, a pronounced interruption of this trend occurs when the inflow is great enough for Mexico to capture more than 74 million cubic meters of water. At that point, Mexican withdrawals into the Acequia Madre Canal stay between 74 million and 76.5 million cubic meters annually, even though U.S. withdrawals continue along their normal trendline. In other words, during high-water years when one would think Mexico should have gotten more water, it didn't.

Two statistical models were developed to test whether (a) the United States drew a fixed share of river water and (b) whether Mexico received a fixed proportion of the river water up to an upper limit (corresponding to the 74 million cubic meter benchmark). Figure 2.2 shows the results of these analyses. It is beyond the scope of this monograph to go into the detail regarding the formulations and validation of these models; technical details are included in the notes to this chapter.

The basic result is that as the volume of water available above El Paso/Ciudad Juárez increases by 100 cubic meters, the U.S. increases its withdrawal on average by 82 cubic meters. As long as Mexico's share is less than the treaty limit of 74 million cubic meters, Mexico increases the water withdrawn from the river by on average 16 cubic meters. However, as soon as Mexico's total supply exceeds 74 million cubic meters, Mexican diversions hit a ceiling.[10] It is unlikely for these observed patterns of water withdrawal to occur on the basis of random fluctuations alone; something is happening.

Has the United States been intentionally holding Mexico's allotment to no more the level called for in the 1906 Convention? Or has Mexico not been taking the water even though it has been available? The answer is: a little of both. From 1939 to 1946, when inflow was above average, the United States simply took the extra water for itself. What was left for Mexico dropped to about 18 percent of inflow (compared with an average of 21 percent), yet the volume released by the United States was still within the limits set by the 1906 Convention.[11] The U.S. practice of not sharing extra water with Mexico came soon after the opening of the American Canal and its diversion dam in

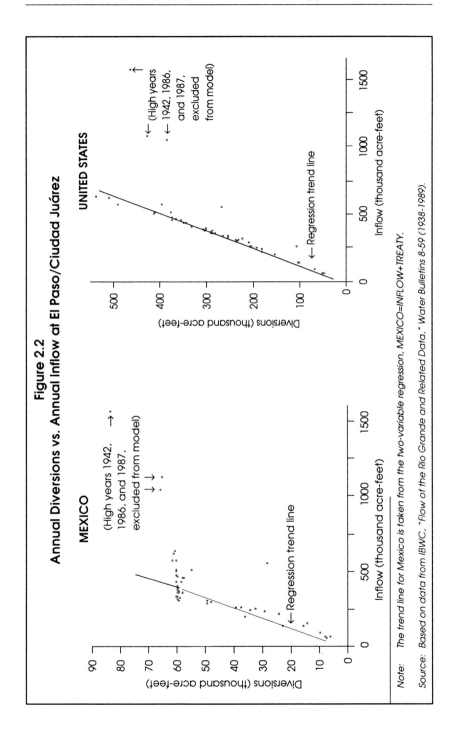

Figure 2.2
Annual Diversions vs. Annual Inflow at El Paso/Ciudad Juárez

Note: The trend line for Mexico is taken from the two-variable regression, MEXICO=INFLOW+TREATY.

Source: Based on data from IBWC, "Flow of the Rio Grande and Related Data," Water Bulletins 8-59 (1938-1989).

1938, so little historical data had accumulated by that time. It also came at the same time the two countries were negotiating a new treaty to allocate the waters of the Rio Grande/Río Bravo below El Paso/Ciudad Juárez and Fort Quitman.

There were also a number of years (1941, 1942, 1981, 1987, and 1988) when natural flows within the Rio Grande/Río Bravo provided Mexico with an opportunity to take more water than it actually did. This is especially apparent for the years 1941, 1968, and 1988. Mexican officials attribute these cases to the fact that the Acequia Madre Canal is old and small; during days of extremely high flow, Mexico has to let water go undiverted because the canal is full to capacity.

U.S. management of water allocation in the river's upper reach appears to be more even-handed now than it was in the early 1940s. Still, the basic problem is that there are too many demands for too little water. Joachin Bustamante, former commissioner for the Mexican Section of the IBWC/CILA, says Mexicans have always resented the fact that the United States takes such a large share of the water even though demands south of the border are just as great.[12] Many experts predict that both shortages and tensions will get worse as the population continues to grow on both sides.[13] But for Mexico to get more water now, either the U.S. Bureau of Reclamation would have to release more water from the reservoirs in New Mexico, or the IBWC would have to divert less water into the American Canal at El Paso. Either option would take away water from current users on the U.S. side who are already demanding more water.

Even if the United States were to practice more equitable apportionment of the river, Mexico would have to expand the capacity of the Acequia Madre Canal in order to capture additional water. Such an investment is unlikely because it would have to compete with many other infrastructure improvements for a share of the limited Mexican public works budget. And unlike wastewater treatment, roads, and sewage systems, the canal's ability to return any social benefits to Mexicans would depend on nature and the good graces of the United States.

At least as pressing in the El Paso/Ciudad Juárez area is the availability of groundwater for municipal and industrial use. Whereas surface water allocation is governed by a bilateral agreement, groundwater use is limited only by the ability of each side to pump it to the surface. The result is that the once-rich aquifers are being depleted so quickly that severe and

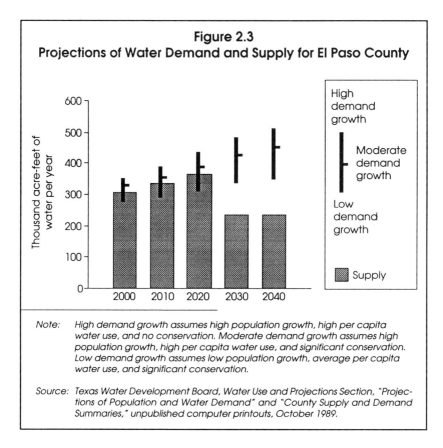

Figure 2.3
Projections of Water Demand and Supply for El Paso County

Note: High demand growth assumes high population growth, high per capita water use, and no conservation. Moderate demand growth assumes high population growth, high per capita water use, and significant conservation. Low demand growth assumes low population growth, average per capita water use, and significant conservation.

Source: Texas Water Development Board, Water Use and Projections Section, "Projections of Population and Water Demand" and "County Supply and Demand Summaries," unpublished computer printouts, October 1989.

chronic shortages are likely to occur within the next 40 years.

Two separate forecasts—one by the TWDB for El Paso County, and the other done by the Universidad Autónoma de Ciudad Juárez for Juárez—indicate shortages before 2030. Both studies probably understate the severity and immediacy of the problem, however, because neither takes into account water use on the other side of the river.

The TWDB projects a significant decline in total water supplies for El Paso County between 2020 and 2030. (See Figure 2.3.) This drop is due almost entirely to an expected reduction in the availability of groundwater. The TWDB expects the supply of groundwater to peak at 253 million cubic meters per year in 2020, dropping severely to 74 million cubic meters per year by 2030. The shortages will have the greatest impact on municipal users (including electric utilities), who currently account for

88 percent of groundwater consumption in El Paso County and who rely on aquifers for about 84 percent of their water.

The City of El Paso has already begun to implement conservation measures to delay future groundwater shortages. The program has demonstrated good results. Although total groundwater use was increasing at an annual rate of 1.4 percent in the mid-1980s, per capita consumption was actually falling by about 1 percent and had reached 2,500 cubic meters per person per year by 1985.

El Paso's municipal water conservation success has been more than offset by growth in Ciudad Juárez. Withdrawals are increasing by about 5 percent every year as the city tries to keep pace with its rapidly expanding population.[14] But there is an even more pertinent trend contained within that figure: *per-capita* consumption, now about half of what it is in El Paso, is increasing at an annual rate of 1.8 percent.[15] Individuals tend to use more water as their living standards improve, and employment in the area's maquiladora industries has led to higher personal incomes in Ciudad Juárez. As the income gap between it and El Paso grows smaller, Ciudad Juárez will continue to face very strong pressure to pump more groundwater.

Both cities take water from the same aquifer system, so the speed with which Ciudad Juárez extracts water will affect the amount of water available for El Paso. And as the amount of sweet water diminishes, an increasing amount of saline water will encroach into the aquifer from surrounding rock formations. In other words, not only will there be less water, it will be less usable.

The issue of groundwater management will be examined more closely in the next chapter. The important point here is that groundwater shortages may occur much sooner than the 2020 predicted by the TWDB, considering (a) the two cities share the same aquifer system, (b) both population and per-capita groundwater consumption are increasing faster in Ciudad Juárez than in El Paso, and (c) the TWDB model does not take into account water use in Mexico.

The Middle and Lower Basin

The 1944 treaty allocates surface water resources in the middle and lower reaches of the Rio Grande/Río Bravo, from Fort

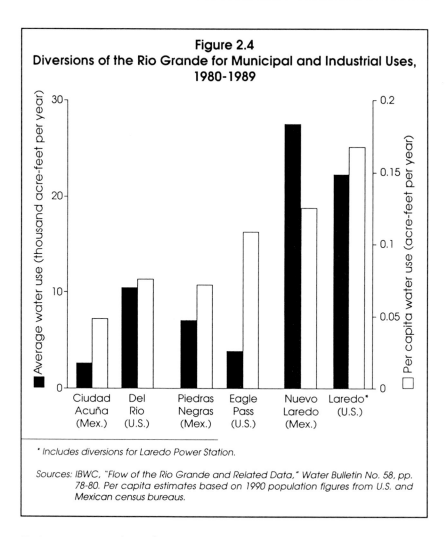

Figure 2.4
Diversions of the Rio Grande for Municipal and Industrial Uses, 1980-1989

* Includes diversions for Laredo Power Station.

Sources: IBWC, "Flow of the Rio Grande and Related Data," Water Bulletin No. 58, pp. 78-80. Per capita estimates based on 1990 population figures from U.S. and Mexican census bureaus.

Quitman in Hudspeth County to the Gulf of Mexico. From Fort Quitman to Falcon International Reservoir, the United States is entitled to all flow reaching the Rio Grande/Río Bravo from the tributaries on its side of the border: the Pecos and Devils Rivers, as well as Terlingua Creek, San Felipe Creek, and Pinto Creek. The treaty also entitles the United States to one-third of the flow from the Mexican tributaries (Río Conchos, Arroyo De Las Vacas, Río San Rodrigo, Río Escondido, and Río Salado), with a minimum of 431.5 million cubic meters irrespective of the total flow.[16] Each country receives half of the flow in the main channel of the Rio Grande/Río Bravo not otherwise allocated. The water in Amistad International Reservoir, located just downstream from the confluence of the Pecos River with the Rio Grande/Río Bravo, is apportioned 56.2 percent to the United States and 43.8 percent to Mexico.[17]

Water is apportioned differently in the lower reaches, from Falcon Reservoir to the Gulf of Mexico. Mexico is entitled to 41.4 percent of water in Falcon Reservoir, with the remaining 58.6 percent designated for the United States.[18] All the major tributaries downstream from the reservoir are on the Mexican side of the border, and Mexico keeps the rights to 100 percent of their flow. The largest of these are the Río Alamo and the Río San Juan in Tamaulipas, Mexico. Mexico and the United States each receive half of the flow in the main channel of the Rio Grande/Río Bravo below Falcon International Reservoir that is not otherwise allocated.

In the middle reach, from Fort Quitman to Falcon Reservoir, Mexican and American withdrawals for municipal/industrial use are both on the order of 45.6 million cubic meters per year.[19] Per-capita use is about 50 percent greater on the U.S. side, however. (See Figure 2.4.) There are no major irrigation diversions by Mexico, while the United States diverts an average of 1.1 billion cubic meters of water per year into the Maverick Canal near Del Rio.

Despite the fact that there are more people (and thus greater demand for municipal and industrial water supplies) on the Mexican side of the border, the 1944 treaty allocates to the United States one-third of the water in the Mexican tributaries in the area between Nuevo Laredo and Ciudad Acuña.[20] The rationale is that the United States is effectively the lower riparian in this area. Below Fort Quitman, most of the flow into the main channel of the river comes from Mexico's tributaries. Only the Pecos and Devils Rivers provide consistent flow from the U.S. side here, since the creeks are intermittent and unreliable.

The major metropolitan area in the middle reach river is Laredo/Nuevo Laredo. The local economy on the U.S. side is heavily steeped in services and trade, with 4.7 percent of personal income derived from manufacturing in food processing, apparel, and printing. Trade with the border regions is expected to play an increasingly important part in the local economy, especially in the event of a North American Free Trade Agreement. Population growth is taking place mostly on the U.S. side of the river, with Laredo growing at an annual rate of 3 percent. Of all the major metropolitan areas along the river, Nuevo Laredo is the only city with a growth rate of less than 1 percent per year.

About 98 percent of all Laredo's water needs—municipal and industrial as well as agricultural—are currently supplied by surface water. The TWDB expects that despite increasing demands, Laredo will be able to find enough water to meet municipal needs throughout the first half of the century. (See Figure 2.5.) This could be accomplished by conservation and by purchasing water rights from agricultural users upstream. But even though municipal needs might be satisfied, the acquisition of water rights by municipalities is expected to contribute to the irrigation shortfall predicted by the TWDB for 2040.[21]

The section of the river below Falcon Dam is the most densely populated and uses the most water. As in the upper reach, diversions from the river in this section go primarily to agriculture. Mexico relies on the Anzalduas Channel Dam near Hidalgo and Reynosa to divert the bulk of its appropriation to an irrigation canal system. Both Mexico and the United States divert about the same amount of water from the river.

Little information is available on municipal withdrawals in the lower Rio Grande Valley in Mexico. According to the IBWC, only two small municipalities directly use the river in this region, and their annual use averaged a mere 1.9 cubic meters from 1978 to 1987.[22] Other Mexican cities may get their public water supplies from the Río Alamo and Río San Juan, however, as Mexico does not need to supply the United States with information on its two tributary flows in this region. In fact, Mexico's use of water stored in reservoirs built on tributaries is not reflected in main channel withdrawal data because the two Mexican tributaries are fully allotted to Mexico. Conflicts may arise in the future, however, if Mexico proceeds with plans to build a municipal reservoir on the Río San Juan to supply water to Monterrey, one of the three largest cities in Mexico.

There are two major metropolitan areas in the lower Rio Grande/Río Bravo: McAllen/Reynosa and Brownsville/Matamoros. The larger of the two—McAllen/Reynosa—is driven economically by trade and services, although in the near future agriculture could take on increasing importance. The McAllen area (which includes Edinburgh) is relatively less reliant on industry than other population centers along the river are. Manu-

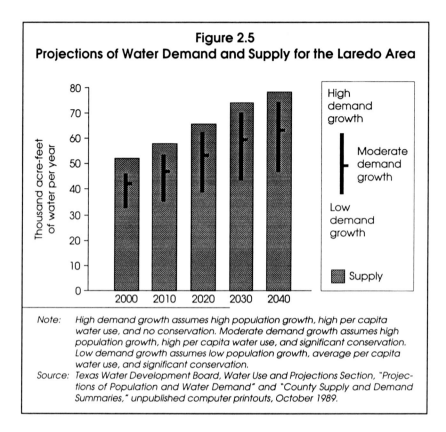

Figure 2.5
Projections of Water Demand and Supply for the Laredo Area

Thousand acre-feet of water per year

High demand growth

Moderate demand growth

Low demand growth

Supply

Note: High demand growth assumes high population growth, high per capita water use, and no conservation. Moderate demand growth assumes high population growth, high per capita water use, and significant conservation. Low demand growth assumes low population growth, average per capita water use, and significant conservation.

Source: Texas Water Development Board, Water Use and Projections Section, "Projections of Population and Water Demand" and "County Supply and Demand Summaries," unpublished computer printouts, October 1989.

facturing in food processing and textiles accounts for 7.4 percent of the personal income of the region. McAllen's population is growing at an annual rate of about 3 percent; Reynosa's growth rate is about 2.5 percent.

Without conservation measures, water quantity problems could be severe in this area. Currently, 92 percent of McAllen's municipal water needs are met by surface water. The TDWC expects water supply to grow slightly during the first half of the century, but not as fast as demand. (See Figure 2.6.) McAllen's population is growing by 3 percent annually, while Reynosa is growing by 2.5 percent. And as with Ciudad Juárez upstream, Reynosa's per-capita consumption of water can be expected to increase as incomes rise.

Quantity problems in the Brownsville/Matamoros area are similar to those faced by McAllen/Reynosa. Groundwater is generally not a feasible alternative to surface water because the

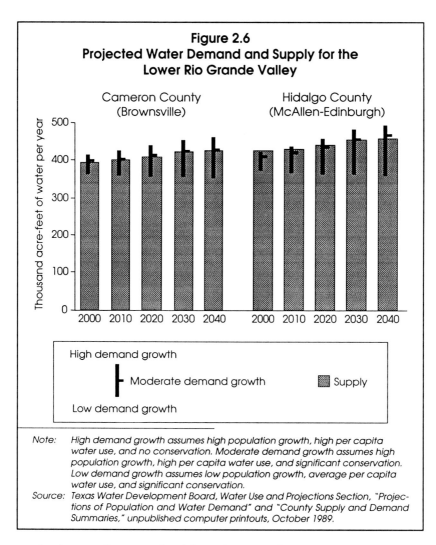

Figure 2.6
Projected Water Demand and Supply for the Lower Rio Grande Valley

Note: High demand growth assumes high population growth, high per capita water use, and no conservation. Moderate demand growth assumes high population growth, high per capita water use, and significant conservation. Low demand growth assumes low population growth, average per capita water use, and significant conservation.

Source: Texas Water Development Board, Water Use and Projections Section, "Projections of Population and Water Demand" and "County Supply and Demand Summaries," unpublished computer printouts, October 1989.

region's aquifers are highly saline, recharge slowly, and are susceptible to subsidence.

Brownsville and Matamoros rely heavily on agriculture, services, and trade. Manufacturing is a significant aspect of the local economy, accounting for 12 percent of personal income. Growth is expected to continue in agriculture, and there is potential for significant growth in industrial capacity. The combined population of the two cities was about 563,500 in 1990, and both are growing at a rate of around 2.3 percent.

Potential Conflicts and Solutions

This overview of water apportionment issues in the Rio Grande/Río Bravo Basin points to some potential conflicts in the future. One of these is the allotment of 74 million cubic meters per year to Mexico in the Juárez area, and the provision in the 1906 agreement which allows the United States to reduce the amount supplied to Mexico in times of drought. Surface water is used for irrigation in the Juárez area, and limiting it prevents expanded use of arable lands in Mexico. As previously mentioned, Mexico historically has not received much additional water during years of relative plenty, although the United States has proportionally reduced Mexico's allotment when total flow falls below average. In addition, Mexico has been limited by the capacity of its own canal.

Shifting water allocations between the nations or renegotiating the existing treaty seems unlikely as a means for resolving these shortages at El Paso/Ciudad Juárez. Mexico is in a poor position to surrender any of its entitlement since it already uses less water than its riparian neighbor. It could claim that any further imbalance in favor of the United States would erode Mexico's already poor living conditions in the border area. On the other side of the table, the United States would face significant internal pressure from farmers, municipalities, and industrial users of domestic water if their allocations were reduced so that Mexico could use more of the Rio Grande/Río Bravo. Moreover, negotiating quantities of water to be swapped internationally could not be done without renegotiating the two treaties that for most of this century have set the rules by which Mexico and the United States share the river. Taken together, the countervailing factors add up to a probable political stalemate between the two countries. And while negotiations dragged on, the population on both sides of the border would continue to grow.

There do exist relatively low-cost, feasible, and practical means to resolve water shortages at El Paso/Ciudad Juárez without changing any of the treaty terms. Table 2.1 lists eight such approaches. Each of these options can be implemented by the existing water institutions of the two nations with no compromise in sovereignty. (Major capital projects, such as new dams or water desalination plants that augment supplies are also feasible, but they are beyond the scope of the marginal improvements listed in Table 2.1.)

The IBWC/CILA has been a success story because the infrastructure it initiated, beginning with the Falcon Dam, made both Mexico and the United States better off. This principle can be extended to three marginal capital improvements:

1. Jointly capture flood waters of the upper Rio Grande to use those waters during low-flow periods.
2. Enlarge and improve the existing Mexican withdrawal canal.
3. Reduce water losses from the existing water transmission and distribution systems.

Figure 2.1 provides the water resources rationale for using existing reservoir capacity at the Elephant Butte Dam or other sites to store flood waters in the upper Rio Grande. In 1986 and 1987 (two of the three highest flow years on record) more than half the flow was undiverted either by Texas or Mexico—more than 715 million cubic feet of water for each of the two years. By comparison, the most water both countries have ever used in any single year is about 740 million cubic meters; the mean is about 400 million cubic meters. Such excess water, if saved and distributed over low-low periods, could add water not now available to the system that would benefit both nations.

Table 2.1
Policy Options for Resolving Water Shortages
at El Paso/Ciudad Juárez

A. **Supply-side alternatives**
Storage to capture under-utilized flood flows
Improve and enlarge the Acequia Madre
Reduce water losses from the existing system

B. **Operational changes in managing the existing system**
Regularize water releases below the American Dam
Inter-connect existing users
Conjunctive use of surface and groundwaters

C. **Demand-side changes to improve water use patterns**
Water conservation—use less water to achieve existing purposes
Shift water to more efficient or desirable uses within each country

---

It is beyond the scope of this monograph to provide the engineering studies needed to demonstrate whether different operating rules would permit existing reservoirs to store floodwaters over-season, but it is recommended that such research be conducted as soon as possible.

As indicated in the previous section, Mexico's Acequia Madre canal is one factor limiting Mexico to slightly more than 74 million cubic meters of water withdrawals from the Rio Grande, even in years when more water is available naturally. A second infrastructure improvement would be to enlarge and improve the canal so that more water could be withdrawn. An improved canal would benefit Mexico by enlarging the volume of water that could be withdrawn, water that could support the thirst of growing municipalities, industries and farms along the Mexican side of the river. An improved canal is no threat to American water users, as the United States controls fully water allocation to its citizens and determines what residual water flows to Mexico.

Reduction of system losses through leak-control and prevention of percolation or evaporation would provide more water to users without augmenting nature's supplies. It is beyond the scope of this monograph to conduct the engineering studies to document whether one-third or one-half of the water withdrawn from the Rio Grande is lost prior to its use (in the field, in industry, or in homes) due to evapotranspiration, fugitive loss, or percolation. However, it is likely that technical investments in the existing water transmission and distribution systems could reduce losses within that order of magnitude. Again, it is recommended that such studies be conducted in the near future.

Better operating policies could enlarge the water available at El Paso and Ciudad Juárez. These options include:

1. Standardizing water releases below the American Dam.
2. Interconnecting existing water users to improve supply reliability.
3. Managing conjunctively existing surface water allocations and groundwater in each nation.

One way to deliver more irrigation water to Mexican growers near Ciudad Juárez would be for the IBWC to develop better release rules for water discharge and withdrawal. The key constraint is that the Acequia Madre has limited capacity. The volume pushed through the canal during the course of an entire year would increase if three reservoir operating rules were adopted. These include:

1. Operating the upstream dams (Elephant Butte Dam in New Mexico as well as the American Dam at El Paso) to provide more water to Mexico at the beginning and end of the wet season (March and September) when the Mexican canal has unused capacity.
2. Increasing flows to the United States by an equivalent amount in June and July when the Mexican canal is often full.
3. During the wet season (April to August) reducing daily and weekly fluctuations in the water released to Mexico to stabilize throughput and avoid capacity limits on Mexican withdrawals.

If these reservoir operating rules were adopted, the volume withdrawn by U.S. users would not change. Mexican users, however, would capture additional water out of flows that previously discharged through the river towards the gulf. It is beyond the scope of this monograph to conduct the analyses to show whether these revised operating rules are feasible from an engineering point of view, are politically acceptable to both sides of the border, or can be justified by patterns of crop water use. But such studies would be the logical next step in defining strategies to improve surface water management.

The El Paso/Ciudad Juárez region is blessed with many separate institutions providing water to farms, factories, and urban and unincorporated areas; each is jealous of its sources. It is beyond the scope of this monograph to demonstrate the truth of an old axiom in the water and energy business: that interconnection of utilities increases the reliability of all supplies. The logic is that a water purveyor can only sell water that it can provide reliably; reliability is affected by the temporal pattern of renewal of its unique set of water sources and its water storage capacity. It is often the case that water purveyors who draw from groundwater, riparian river flow, or controlled reservoir releases can each improve significantly their supply reliability while simultaneously reducing significantly their required off-line storage capacity by being able to rely on inter-connections that allow firms to take advantage of nature's differences in the volume of water available over time and space.

This principle of diversity of supplies is illustrated also by the advantages of conjunctive management of surface and groundwater in the El Paso/Ciudad Juárez region. An aquifer can be viewed as an underground reservoir that has several advantages over surface water storage: percolation and evaporation are limited, water quality can be improved through soil migration, and water can be withdrawn closer to the point of use. Excess flood flows could be diverted to recharge the regional aquifers. Managing both surface and ground supplies together allows water utilities to provide water at a lower cost with more security. But since groundwater supplies are also dwindling (see Chapter 3), it is important that conjunctive use take place in the context of comprehensive watershed management. The next chapter provides a more detailed discussion of groundwater use, and offers suggestions as to how a bilateral strategy of comprehensive watershed management can be successful.

Changes in the demand for water could provide "new" supplies even more cost-effectively than the supply augmentation or improved water management alternatives discussed above. Two approaches that can be implemented unilaterally by the Americans or Mexicans without consultation or cooperation are water conservation and water reallocation.

It is possible to use less water to accomplish the same purposes in any sector: domestic, industrial or agricultural. For example, the Texas Water Commission estimates that on the order of 20 percent of domestic water supplies in Texas could be saved and reallocated to other uses through available, cost-effective water conservation steps.[23] Crops on both sides of the border can be irrigated with less water with shifts from current to water-conserving technologies, such as sprinklers or drip irrigation. It is beyond the scope of this monograph to identify the specific demand-side benefits that could be achieved in the El Paso/Ciudad Juárez region.

Water rights can be shifted voluntarily by U.S. users on the Texas side of the border and among Mexicans. Sales of water or water rights provide new water to some and income in lieu of flow to others. Indeed, below the Falcon Reservoir on the Rio Grande/Río Bravo there has been a shift of over 34.5 million cubic meters of water rights from agricultural to municipal uses.[24] On the U.S. side the sales of water or transfers of water rights can be conducted through the Rio Grande River Master's office. It is beyond the scope of this monograph to report the scale of potential voluntary shifts of water or water rights among users of the El Paso/Ciudad Juárez region.

The uncertainty of water demand and supply into the next century emerges as a potential source of conflict between both the United States and Mexico and between municipal, industrial, and agricultural users on either side of the border. It is clear that Mexico's per-capita water consumption is likely to increase as living standards increase and as cities on both sides of the border grow. Although many policy options exist to manage the bilateral tensions over available water, the no-action option is the one that has provided the basis for the fear of Mexican Ambassador Cesar Sepulveda when he wrote:

> One of the questions that can most affect diplomatic relations between Mexico and the United States in the latter part of the 20th century, if not corrected, will be the theme of the water resources shared by the two countries in the frontier area.[25]

Notes

1. Complete data for this time are unavailable. However, partial data maintained by the U.S. Geological Survey indicate that for 1901 and 1902, the combined U.S. and Mexican consumption of surface water from the Rio Grande/Río Bravo amounted to only one-third of the available flow. This is comparing the flow of the river just above El Paso/Ciudad Juárez with the flow 40 miles downstream near Ft. Hancock. U.S. Geological Survey, "Water Resources of the Rio Grande Basin 1888-1913," Water Supply Paper No. 358, Washington, 1915. For population estimates, see *The Texas Almanac, Dallas Morning News,* 1925.

2. Texas Water Development Board, *Water for Texas: Today and Tomorrow, 1990,* December 1990.

3. International Boundary and Water Commission, "Flow of the Rio Grande and Related Data," Water Bulletin No. 58 (1988). See tables for relevant diversions.

4. TWDB, *Water for Texas.*

5. TWBD, Water Use and Projections Section, "Projections of Population and Water Demand" and "County Supply and Demand Summaries," unpublished computer printouts, October 1989. See also TDWR, Water Use, Projected Water Requirements, and Related Data and Information for the Metropolitan Statistical Areas in Texas, Austin, Texas, July 1985, pp. 44-48, 72-76, 107-110, 125-129.

6. Convention between Mexico and the United States for the Equitable Division of the Waters of the Rio Grande for Irrigation Purposes, Articles I and II, 1906.

7. U.S. Geological Survey, "Surface Water Supply of the United States," Water Survey Paper 588, 1928; International Boundary Commission, "Flow of the Rio Grande and Related Data," Water Bulletin Nos. 1-9, (1931-1939).

8. Based on average annual flow data as reported by the IBWC in "Flow of the Rio Grande and Related Data," Water Bulletin Nos. 9–59 (1939-1989). Data set included all years from 1939 to 1989 except flood years of 1942, 1986, and 1987. Standard deviations for the portion diverted by the United States in any given year was 6.5 percent, and by Mexico 2.7 percent.

9. Calculated from average monthly flows for April through August. See IBWC, "Flow of the Rio Grande and Related Data," Water Bulletin No. 59 (1989), pp. 10-12.

10. The authors used a single-variable, ordinary least-squares regression to model U.S. withdrawals against available inflow:

$$\text{TO US} = \alpha + \text{ßINFLOW} + u.$$

Both variables represented annual flow volumes for any given year between 1939 and 1991. The results were:

Adj. R^2:	.97
F Ratio:	1536.17
Observations:	48
Durbin-Watson:	1.996

Parameter	Estimate	t Ratio	Prob>|t|
INFLOW	.82	39.2	0.0001

To model Mexican withdrawals and to test whether 60,000 acre-feet was a significant turning point in withdrawal patterns, a piecewise generalized least-squares model was used:

$$\text{TO MEXICO} = \alpha + \beta_1 \text{INFLOW} + \beta_2 \text{OVER60K} + u.$$

OVER60K was a dummy variable equal to 0 if TO MEXICO was less than 60,000; otherwise it was equal to TO MEXICO – 60,000. The results were:

Adj. R^2:	.84
F Ratio:	107.88
Observations:	41
Durbin-Watson:	1.89

Parameter	Estimate	t Ratio	Prob>|t|
INFLOW	.16	13.96	0.0000
TREATY	-5.38	-1.32	0.195

Data were taken from the IBWC, "Flow of the Rio Grande and Related Data," Water Bulletins Nos. 9–61 (1939-1991). Data for the years 1942, 1986, and 1987 were excluded from the model because runoff for those years was unusually high—853 percent, 651 percent, and 505 percent of normal. Data for 1990 were excluded because a diesel spill forced the closure of the American Canal for most of that year. Because these necessary exclusions made the adjusted values misleading, data for the years 1939-1943 and 1985-1991 were excluded from the GLS model.

11. This excludes the flood year 1942.

12. Joachin Bustamante, telephone conversation with David Hurlbut, June 12, 1992.

13. See Milton H. Jamail and Stephen P. Mumme, "The International Boundary and Water Commission as a Conflict Management Agency in the U.S.-Mexico Borderlands," in *The Social Science Journal,* Vol. 19, No. 1 (January 1982), pp. 45-62.

14. María del Rosario Díaz A. and Alfredo Granados Olívas, "Evaluación Geohidrológica del Acuifero de la Zona Urbana de Ciudad Juárez, Chihuahua, Período 1980-90," unpublished paper, Instituto de Ingeniería y Arquitectura de la Universidad Autónoma de Ciudad Juárez, September 11, 1991.

15. Calculated using data from del Rosario Díaz and Granados, and the Mexican Census Bureau.

16. United Nations, *Treaty Series,* New York, 1947, pp. 315-358.

17. Ken Rakestraw, Water Accounting, IBWC, El Paso, Texas. Phone interview, April 25, 1990.

18. Ken Rakestraw interview.

19. IBWC, "Flow of the Rio Grande and Related Data," Water Bulletin No. 59 (1989).

20. The treaty alternatively specifies a minimum flow of 431.5 million cubic meters to be available to the United States. In practice, the IBWC concerns itself with ensuring this a five-year average is received through flow gauges located at the confluence of these Mexican tributaries. Ken Rakestraw interview.

21. William Moltz, Texas Water Development Board, interview with David Hurlbut, June 24, 1992.

22. IBWC, "Flow of the Rio Grande and Related Data," Water Bulletin No. 57, pp. 78-80.

23. In its 1991 Strategic Plan for Texas, the Water Commission established a goal of achieving a 20 percent reduction in water use by the year 2000. Personal communication from Mike Personett of the Texas Water Commission to David Eaton, August 25, 1991.

24. F. Andrew Schoolmaster, "Water Marketing and Water Rights Transfers in the Lower Rio Grande Valley, Texas," *The Professional Geographer,* Vol. 43, 1991, pp. 292-304.

25. Ambassador Cesar Sepulveda, "U.S.-Mexican Transboundary Needs and Issues to the Year 2000," working group report, Corpus Christi, Texas, 1982, pp. 7-8.

Chapter 3:
Groundwater

G ROUNDWATER MANAGEMENT IN THE RIO GRANDE/
Río Bravo Basin poses particularly difficult policy challenges because of the technical uncertainties and lack of
control by both nations. The quality and quantity of groundwater in the border area's high-growth centers are already diminishing, and without systematic planning and intervention, the
population pressures that accompany economic integration
between Mexico and the United States cannot but exacerbate
future water shortages.

Texas and Mexico have no history of cooperation to protect
or share their aquifers. No present institution or treaty between
the two countries deals with issues affecting international
groundwater supplies. Compounding the problem is that the
two countries have different domestic laws governing groundwater use. Any agreement that is fair bilaterally could require
one or both nations to strike down state laws and local precedent for allocating withdrawal rights, which might not be politically feasible.

This chapter explores the politics of scarcity as it pertains to
economic development and groundwater resources in the Rio
Grande/Río Bravo basin. It will begin by describing nature's view
of the problem: how an aquifer works, how it can be damaged,
and the current state of the shared aquifers lying below the international border area. It will then examine the current policies governing groundwater use in Mexico and in Texas, policies
which were devised and carried out without regard to their long-
term impact on water availability along the border. The discussion will then examine groundwater issues that may be raised
in transboundary environmental policy negotiations and will
offer suggestions as to how a treaty may address these issues.

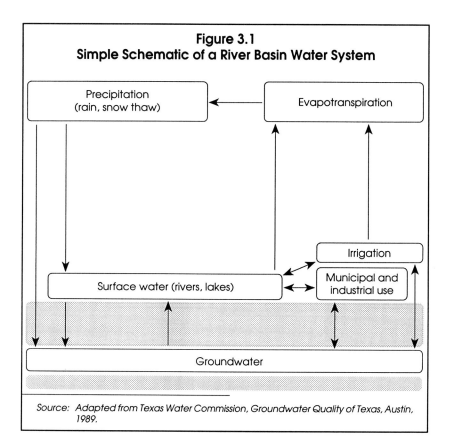

Figure 3.1
Simple Schematic of a River Basin Water System

Source: Adapted from Texas Water Commission, Groundwater Quality of Texas, Austin, 1989.

Aquifers of the Rio Grande/Río Bravo Basin

Aquifers are an integral component of a river basin's water cycle. They are geological formations below the earth's surface with low-density, high-porosity rock capable of absorbing and storing water. Percolation from rainfall and from the river bed replenishes underlying aquifers, while springs and wells return water to the surface. Consequently, the way the surface flow is diverted, depleted, or polluted can interfere with the ability of the river system to recharge its groundwater reserves. The natural factors affecting the recharge and discharge of groundwater include soil porosity, the slope of non-porous rock strata underlying water-permeable strata, rainfall patterns, and fluctuations in the volume of surface water in an adjacent river or lake.

Figure 3.1 shows how an aquifer is linked to the rest of a river basin's water system. Rainfall that does not enter a stream can be absorbed by the soil and then pass down through permeable rock below the surface. Rivers and streams also contribute to groundwater as they flow across rock formations, intersecting the river bed which contains water-permeable formations. Once underground, the water flows downhill from the recharge area along the contours of impermeable bedrock. Occasionally, water will collect in depressions formed by the stratum of non-porous rock; these can usually be tapped by wells. The porous strata can also re-emerge at points on the surface, creating springs and lakes.

Groundwater dissolves and carries with it the various minerals found in the soils through which it passes. Consequently, groundwater farther downhill will contain more dissolved solids than water nearer the recharge area. The chemicals collected by the groundwater can be natural (such as chlorides, nitrates, and other dissolved ions) or the result of human activity (pesticides, brine from oil and gas wells, leakage of industrial wastes, or nutrient runoff from fertilizer applications).

Since aquifers are not readily visible like surface water bodies, it is difficult to map them with certainty. Precise migration patterns of groundwater are not easily tracked, nor can the extent of contamination be gauged reliably beyond the points of measurement.[1] Still, the general model for understanding groundwater behavior is straightforward:

If "water out" is greater than "water in," the difference comes out of the aquifer's accumulated storage. If water mining is sustained over a long period of time, recoverable water can be depleted or polluted, sometimes causing permanent damage to the aquifer's ability to recharge and supply potable water.

Table 3.1 shows the best available data for the major aquifers along the Rio Grande/Río Bravo. These include the Hueco and Mesilla Bolsons in near El Paso/Ciudad Juárez, the Red Light Draw Bolson, the Green River Valley Bolson, the Presidio and Redford Bolsons, the Edwards-Trinity Aquifer, the Carrizo-Wilcox Aquifer, and the Gulf Coast Aquifer.

Table 3.1
Estimates of Groundwater Supplies and
Withdrawals in the Rio Grande-Río Bravo Basin
(thousands of acre-feet)

Aquifer	Annual recharge	Recoverable storage 1980	Projected annual groundwater withdrawals				
			1990	2000	2010	2020	2030
Alluvium and Bolson (El Paso/Ciudad Juárez to Presidio/Ojinaga)	434.0	32,265.5	952.1	989.7	1,027.5	1,016.9	469.9
Edwards-Trinity (Del Rio/Ciudad Acuña)	776.0	n/a	776.0	776.0	776.0	776.0	776.0
Carrizo-Wilcox (Laredo/Nuevo Laredo)	644.9	9,909.2	828.7	828.7	828.7	828.7	644.9
Gulf Coast (McAllen/Reynosa to gulf)	1,229.8	n/a	1,229.8	1,229.8	1,229.8	1,229.8	1,229.8

Source: *Texas Water Commission,* The State of Texas Water Quality Inventory, *9th Edition, 1988.*

The left two columns in Table 3.1 are estimates of the amount of available water: annual recharge plus current recoverable storage. The right five columns are estimates of how much groundwater could be drawn from these aquifers each year up to 2030, assuming current withdrawal patterns. Note in particular the pattern for the Alluvium and Bolson formations from El Paso/Ciudad Juárez to Presidio/Ojinaga. Estimated withdrawals from the aquifers are twice the recharge rate, which consequently will deplete the recoverable storage over time (see Figure 3.2). If these figures are accurate, by 2030 the aquifers will be able to supply only half of what they are supplying now, since little will be available from accumulated storage. As mentioned in the previous chapter, the Texas Water Development Board predicts a sharp decline in the amount of water withdrawn from the El Paso aquifers—from 253 million cubic meters per year in 2020 to about 74 million cubic meters in 2030. Behind this decline is a reduction in the amount of recoverable storage in the aquifers. The Texas Water Commission estimates that if current patterns continue, the total recoverable storage in these

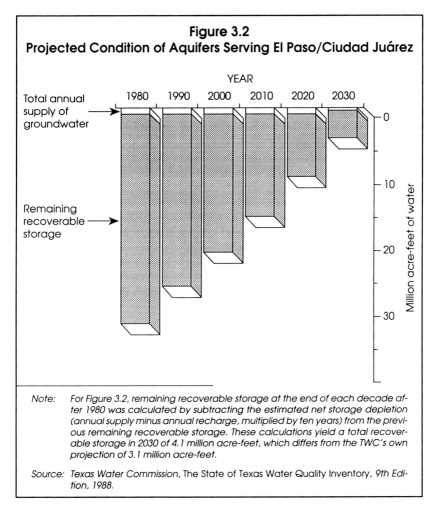

Figure 3.2
Projected Condition of Aquifers Serving El Paso/Ciudad Juárez

Note: For Figure 3.2, remaining recoverable storage at the end of each decade after 1980 was calculated by subtracting the estimated net storage depletion (annual supply minus annual recharge, multiplied by ten years) from the previous remaining recoverable storage. These calculations yield a total recoverable storage in 2030 of 4.1 million acre-feet, which differs from the TWC's own projection of 3.1 million acre-feet.

Source: Texas Water Commission, The State of Texas Water Quality Inventory, 9th Edition, 1988.

aquifers will dwindle from the current 39.5 billion cubic meters to just over 3.7 billion cubic meters by 2030.

The quality of much of the groundwater along the Texas/ Mexico border is poor by Texas Department of Health standards.[2] Salinity is a common problem, particularly in deep aquifers. The alluvium and bolson deposits tend to have a relatively high natural salinity, as does the Gulf Coast Aquifer. The natural salinity of any groundwater source is exacerbated when water is withdrawn faster than the aquifer can recharge. Human-caused groundwater degradation is related to problems of municipal infrastructure (inadequate wastewater treatment, leaching from city land fills), poor regulation of industrial waste disposal (including hazardous waste), agricultural practices, petrochemical and mining activities, leaks and spills, and abandoned or faulty water wells.

Lack of sewage treatment facilities is a problem for many cities on the Mexican side of the border and for low-income colonias on the U.S. side. Often, raw sewage is dumped into the river or onto the ground without treatment, contaminating the aquifer for both sides of the border. The Carrizo-Wilcox and Gulf Coast aquifers, where gravel outcropping is near the ground's surface, are particularly vulnerable to contamination caused by dumping of untreated sewage, the use of septic tanks, and the overflow caused by periods of intense rainfall. Pathogenic microorganisms can enter the groundwater supply, along with solvents and other chemicals. Chemicals leaching from public landfills can also contaminate an aquifer, although the lack of monitoring facilities makes it difficult to determine the kinds or volumes of chemicals passing into the groundwater.

Hazardous waste is the greatest health-threatening form of groundwater degradation, and this problem has increased in tandem with the border's economic growth. Tests at six Texas locations designated by the EPA for hazardous waste clean-up have shown groundwater contamination at shallow levels. Three of these sites are located along the Rio Grande/Río Bravo: two in Mission, one in Harlingen.[3] The extent of contamination in the Mission aquifer is still uncertain; in Harlingen, the tests found that the shallow upper sand aquifers contained traces of pesticides and arsenic. Officials believe the chemicals in the Harlingen aquifer came from abandoned storage facilities as well as from the regular agricultural practices of farmers and growers in the Lower Rio Grande Valley. But so far, the concentration of chemicals has not reached a level that renders the water in the upper aquifer unsafe for drinking, according to the Texas Department of Health's standards. The quality of the lower aquifer is uncertain.

Pesticides enter the groundwater system the same way natural salt does. As surface water passes downward through the porous soils, chemicals dissolve and are carried by the water into the aquifer. Along the Rio Grande/Río Bravo, the greatest potential for pesticide contamination is near the Brownsville/Matamoros and McAllen/Reynosa areas, where agricultural activity is intense on both sides of the border.[4]

Problems in the El Paso/Ciudad Juárez Area

The Hueco and Mesilla Bolson Aquifers that underlie the El Paso/Ciudad Juárez area are composed of unconsolidated beds of gravel, silt, sand, and clay.[5] The Hueco Bolson extends into the state of Chihuahua in Mexico, while the Mesilla Bolson extends into New Mexico. The water quality ranges from fresh to slightly saline. Most of the water that is recovered from the Hueco Bolson aquifer has less than 500 milligrams per liter of total dissolved solids.[6]

As discussed in the previous chapter, El Paso and Ciudad Juárez rely on the Hueco and Mesilla Bolsons for almost all of their municipal water supplies. El Paso withdrew more than 101.1 million cubic meters of groundwater in 1985, while the more populous Ciudad Juárez withdrew about 82.6 million cubic meters. Since then, however, total groundwater withdrawals in Ciudad Juárez have increased 6.8 percent annually, due both to an increasing population and increasing per-capita water consumption. The Texas Water Commission estimates that withdrawals are increasing by 2.2 percent annually in El Paso County. Population is growing more slowly in El Paso (2.1 percent annually as compared with 3.5 percent in Ciudad Juárez) while per-capita water consumption is decreasing by about 1 percent annually.

The recharge of both aquifers is significantly less than the withdrawals. The Mesilla and Hueco Bolsons together have an annual natural recharge of 29.6 million cubic meters—about 16 percent of what the twin cities withdrew in 1985. A 1982 study projected that accelerated overdrafting in recent years would cause the groundwater storage level in the Hueco Bolson to decline as much from 1973 to 1990 as it did for the prior seven decades.[7] The Texas Water Commission measured the water depth in wells across the state from 1975 to 1985, and during that time the water table in many parts of the El Paso/Ciudad Juárez area fell 6.2 meters to 12.2 meters—in some locations, even more.[8] Similar reductions were recorded at wells in Ciudad Juárez.[9] And as the quantity of water stored in the Mesilla and Hueco Bolsons gets smaller, the salt content of both aquifers becomes more concentrated, making the remaining groundwater progressively less usable.

Pesticides and nitrates contained in irrigation runoff, contaminants leaching into the ground from waste dump sites, and dumping of untreated sewage also pose serious risks to groundwater in the El Paso/Ciudad Juárez region. At least 16 underground storage tanks were leaking contaminants into the ground in El Paso County as of 1988. Such tanks contain gasoline, diesel, or chemicals such as naptha, acetic acid, xylene, toluene, methyl ethyl ketone, and various mixed solvents. The Texas Water Commission also identified 45 oil and chemical spills in 1987 that affected surface and groundwater supplies in El Paso County.[10] A major diesel spill caused the closure of the American Canal in El Paso for most of 1990.[11]

Dumping of untreated sewage, a long-time problem in Ciudad Juárez, allows fecal coliform and other bacteria to seep into the aquifer. These micro-organisms can increase the incidence of cholera, giardia, amoebic dysentery, and other water-borne oral-fecal diseases. In July 1992, cholera bacteria were found in a Ciudad Juárez drainage ditch, raising concern that the disease could be carried to El Paso. Especially at risk are the low-income residents of El Paso's colonias, where sewage treatment is non-existent and drinking water is often contaminated.[12]

Red Light Draw, Green River Valley, Presidio, and Redford Bolsons

Most of the water from these aquifers is of good quality and is used mostly for agricultural purposes: irrigation and livestock wells. Groundwater in these three aquifers flows from north to south.

The Red Light Draw Bolson has a groundwater storage capacity of approximately 740.1 million cubic meters; the annual level of recharge is 2.5 million cubic meters and the recoverable storage is 555.1 million cubic meters.[13] The Red Light Draw Bolson lies between the Eagle and Indio Mountains and the Quitman Mountains. The aquifer is recharged from the mountains, and from there the groundwater migrates from north to southeast towards the Rio Grande. Along most of this distance, the groundwater has less than 500 milligrams per liter of dissolved solids—fresh water usable for domestic purposes. Accumulation of dissolved solids renders the groundwater saline and unfit for drinking or irrigation by the time it reaches the river.

In terms of geohydrology, the Green River Valley, Presidio, and Redford Bolsons are similar to the Red Light Draw Bolson. Like the Alluvium and Bolson Deposits, these aquifers are made of unconsolidated beds of sand, clay, gravel, and caliche.[14] The annual recharge of the Green River Valley Bolson is 1.2 million cubic meters; of the Presidio and Redford Bolsons, 8.6 million cubic meters. (See Table 3.2.)

As with the Red Light Draw Bolson, the accumulation of dissolved solids reduces the water quality as it flows towards the Rio Grande. The water from these aquifers is used primarily for livestock and irrigation. Since fresh water is at higher elevations, there are no problems associated with the use of groundwater from these aquifers, as long as the water levels are kept relatively high. No serious water quality problems currently affect development of these aquifers.[15]

Edwards-Trinity (Plateau), Carrizo-Wilcox, and Gulf Coast Aquifers

The three remaining aquifers (the Edwards-Trinity, Carrizo-Wilcox, and Gulf Coast Aquifers) vary in their general levels of water quality. The Edwards-Trinity zone is a large aquifer with a large potential yield, but not all groundwater is recoverable because of salinity problems. The Edwards-Trinity Aquifer comprises a lower section of fine and coarse-grained sand 30 meters feet thick, and a 300-meter thick upper level of limestone with solution cavities and fractures. This allows water in the Edwards-Trinity Aquifer to travel rather quickly.

The Edwards-Trinity Aquifer serves as the source of water for municipal and industrial uses in the sparsely populated area east of Big Bend National Park. Presidio gets more than 13 percent of its municipal water from this aquifer. It can sustain an average annual withdrawal rate of 185 million to 234.4 million cubic meters. Groundwater withdrawals as of 1980 amounted to around 160.4 million cubic meters annually, primarily for irrigation. In a few heavily irrigated areas, recharge does not keep pace with withdrawals.[16]

Located in the lower middle part of the Rio Grande, the Carrizo-Wilcox Aquifer consists of two layers: a lower layer of lignite beds and a top layer of fine- to coarse-grained sand, clay, silt, and sandstone. The total saturated thickness of the aquifer ranges between 61 meters and 213 meters. Most of the wells in the area have an approximate yield of 1,900 liters per minute, and the water quality ranges from fresh to slightly saline. The annual recharge is 16.9 million cubic meters, with a recoverable storage of 197.4 million cubic meters.

Eagle Pass gets nearly 10 percent of its municipal water supply from this aquifer. Laredo relies on the Carrizo-Wilcox and other aquifers for less than 2 percent of its total water needs. Water from this aquifer is used mostly for irrigation in the Winter Garden zone, although some is used by Angelina and Nacogdoches Counties for municipal and industrial purposes.

It is in the area of the Edwards-Trinity (Plateau) and Carrizo-Wilcox aquifers that the greatest possibility exists for groundwater contamination from petroleum operations. The Texas Water Commission has identified seven border counties in which abandoned oil rig supply wells might exist. But none of the counties along the border have ever been major oil producers, and Texas Railroad Commission has verified few instances of groundwater contamination by abandoned wells.[17]

The Gulf Coast Aquifer in the Brownsville/Matamoros area is too saline to use as an alternative to the surface water. Nonetheless, some farmers occasionally use groundwater despite its high salinity, which has contributed to crop damage in the area. Clay strata between the surface and the underlying aquifer zones tend to trap most of the fertilizer and insecticide particulates that are caught in irrigation runoff, but irrigation is so heavy in the lower Rio Grande/Río Bravo valley that a considerable amount still reaches the aquifers. This and the groundwater's natural salinity render most of the water in the aquifer unsuitable for human consumption.[18]

The annual recharge of the aquifer is estimated at 14.1 million cubic meters, sufficient to sustain the current level of groundwater withdrawals in most areas. Less than three-tenths of a percent of water demand in Brownsville and the rest of Cameron County is met by groundwater; in McAllen and surrounding Hidalgo County, only 3 percent comes from aquifers. Salt water intrusion from the coast especially in the Brownsville/Matamoros area makes groundwater largely unusable for agriculture, industry, or municipalities.

Leaks from underground gasoline storage tanks have contaminated groundwater supplies in the valley at many locations. State officials say that a majority of old tanks that have been excavated have leaked measurable amounts of petroleum products into nearby aquifers. Another possible threat to groundwater supplies is the disposal of industrial waste into underground injection wells. State authorities have granted permits for 118 Class I underground injection wells that inject industrial waste into areas of the aquifer. These injection strata are deep, moderately saline, and have more than 3,000 milligrams of total dissolved solids per liter, a level the state considers not fit for human consumption.[19] So far, no contamination has been associated with these wells.

Groundwater Law

The United States and Mexico have different legal reasons behind groundwater nonmanagement. The Mexican federal government has the authority to enforce a unified plan of conservation and allocation but has not chosen to do so. In the United States, the legal framework is inadequate and institutionally chaotic. Missing from both sides are the principles international jurists say ought to guide international agreements governing shared aquifers.

One of the problems that would be posed to any bilateral plan to manage aquifers within the Rio Grande/Río Bravo basin is the lack of clarity of international groundwater law. The field of international surface water law is well developed, based upon precedents in national and international court decisions, bilateral and multi-lateral treaties, resolutions of international bodies, and statements by experts in international law. Principles for managing international aquifers have been formulated by international bodies and international law experts, but have yet to be tested in wide practice. In 1966 experts developing the Helsinki Rules of international water law declared that groundwater was part of a river system, and thus should be included in water resource negotiations involving international river basins. In 1988, an international conference of government officials and water law experts recommended that in treaty negotiations, parties agree to a "systems approach" to defining water resources. In other words, a treaty ought to begin by recognizing that surface water, geology, rainfall, and aquifers constitute a natural water system with quantifiable inputs and outputs. Parties then should agree on the extent of groundwater resources as well as principles and procedures governing the amount of water to be withdrawn in any country dependent on the aquifer. This would also govern activities in any one country that would affect the quality of groundwater flowing to any other country.[20] The general principles would include:

> the right of each basin State to an equitable utilization
> and the duty not to cause appreciable harm to a co-
> basin State (including to the environment), and recog-
> nize the duty to exchange available relevant
> information and data, the duty to notify and to con-
> sult reciprocally with co-basin States that may be ad-
> versely affected by a project or program planned by
> one or more basin States and the duty to consult con-
> cerning the institutionalization of cooperation or col-
> laboration for basin development upon the request of
> any other basin State.[21]

The doctrine of comprehensive management takes equitable utilization a step further. Under this approach, a bilateral organization would deal with apportionment and management of groundwater resources on the basis of equitable utilization principles.[22] It could execute the kind of systems analysis management methods called for by experts in international law. This is the most bureaucracy-intensive doctrine guiding a bilateral agreement, as it involves either creating a new joint commission or revising the mandate of an existing body such as the International Boundary Water Commission (IBWC) so that it can take on the added responsibility.[23]

The comprehensive management principle is exhibited in detail in the so-called Bellagio draft treaty on transboundary groundwaters.[24] Drafted in 1989 by experts in water management and international law, the document is intended to provide a starting point for the development of a working agreement on groundwater management. Under the Bellagio framework, nations that share aquifers would establish an international commission to maintain data and draw up comprehensive management plans for common groundwater supplies. Details of the management plans would be determined by the commission, while enforcement would be left to the individual nations.

One bilaterial groundwater treaty exists between France and Switzerland that accepts the principles of limited national sovereignty, equitable utilization, data exchange, prior notice and consultation.[25]

Alternative doctrines are those of prior appropriation and

present use. Prior appropriation allocates water based on the idea of earliest beneficial use.[26] For the purposes of a groundwater treaty, the United States would benefit at the expense of Mexico because groundwater uses were established earlier in the United States than in Mexico. If negotiations were to proceed under this doctrine, both parties would have an incentive to increase their beneficial uses, which could lead to competitive and wasteful pumping between both countries.

The most progressive of these doctrines (equitable use and comprehensive management) have never been tested in an international court. There is a precedent in environmental treaties for (1) expanding the authority of existing international institutions to sponsor aggressive technical research, to recommend a plan of action, and to monitor compliance; (2) allowing for a periodic review renegotiation of the specific restrictions imposed by the treaty; and (3) determining whether and how a party's status as a developing country affects issues of finance and compliance. The 1987 Montreal Protocol on substances that deplete the ozone layer uses an existing multilateral organization (the United Nations Environment Program) to coordinate technical research into various aspects of stratospheric ozone degradation. As conditions warrant scientific data are used to review and recommend changes to the protocol. The data and recommended amendments then are presented at a meeting of the parties. This process of ongoing review and revision was incorporated into the protocol to give the agreement enough flexibility to adjust to rapidly improving scientific evidence, technological advances, and economic needs of developing countries.[27]

Any treaty between the United States and Mexico is bound to require some change in domestic groundwater policy in one or both countries. In the United States, groundwater policy is decentralized. There is no national policy; it is left to the states to enforce laws and policies that balance competing local and regional interests.[28] The general practice in Texas specifies that well-owners can withdraw as much as water as they want from beneath their property, even if a well-owner's use of groundwater "cuts off the flow of such waters to adjoining land, and deprives the adjoining owner of their use."[29] This is commonly called the right of capture. Texas neither apportions nor limits groundwater withdrawals as it does with surface water.

Texas law makes a distinction between percolating waters and underground streams. Section 52.001 of the Texas Water Code excludes underground streams from the provisions of groundwater law, although the definitions of these two distinct concepts are still a matter of legal debate. However they may be defined, underground streams flowing below private property belong to the state and may be regulated.[30] In 1992, the Texas Attorney General designated the Edwards Aquifer as an underground river and thus subject to explicit state regulation. This precedent could affect management of groundwater adjacent to the Rio Grande if any of those aquifers are declared underground rivers. In such a case Texas could allocate groundwater withdrawal rights or encourage users to manage the groundwater to assure sustained yields.

Although the common law rule allows unlimited withdrawals of groundwater, well owners do have legal recourse if a neighbor injures their ability to withdraw groundwater, and does so for the sole purpose of inflicting damage. In such cases, common law recognizes the right of the injured party to take legal action against the neighboring well owner. Well owners can also be legally responsible for water that is wasted from an artesian well. In addition, legal action can be taken against industrial firms (such as oil and gas drillers) whose operations contaminate groundwater supplies.[31]

In Mexico, groundwater is property that transfers with land ownership, so owners can drill wells and extract water as needed. However, the Mexican Constitution grants state ownership rights to all subsoil natural resources, including groundwater. Article 27 delineates the government's authority to intervene to limit withdrawals or regulate groundwater use:

> Underground water may be brought to the surface by artificial works and utilized by the surface owner, but if the public interest so requires or use by others is affected, the Federal Executive may regulate its extraction and utilization, and even establish prohibited areas, the same as may be done with other waters in the public domain.[32]

Thus, Mexico in theory treats groundwater as a public good that is subject to government allocation. Groundwater policy is centralized to such an extent that local and regional interests have little leverage over the government's decisions on allocation. Most legal authority in matters of water resources management, from irrigation to groundwater apportionment, is vested in the Secretaría de Agricultura y Recursos Hidráulicos (SARH).[33] Although the Mexican government is implementing a program to decentralize decisionmaking, the process remains concentrated in the federal agencies in Mexico City.

Issues for Negotiation

The two primary issues a groundwater treaty between the United States and Mexico will have to address are apportionment and management. These political and economic questions are dependent on the hydrological constraints imposed by nature. But the paucity of hydrological data on both sides of the border makes it difficult to determine exactly what those natural constraints are. The necessary technical information on groundwater withdrawals, migration patterns, and the location and quantity of groundwater resources is generally unreliable or unavailable. Even in those areas such as the Upper Rio Grande for which information is more readily available, the geohydrological conditions are still a matter of debate among experts.[34] On the Mexican side, it is very difficult to get any data on groundwater withdrawals; it is possible that SARH has not undertaken studies to gather this data. The IBWC, which has collected extensive data on surface water, does not have the mandate to resolve groundwater issues and thus has only limited data on aquifers.

A number of other specific issues exist which will have to be addressed directly. One is the problem of groundwater contamination due to poor storage of industrial wastes. This is a trade-related issue, although it has never been addressed explicitly in the context of trade negotiations. The elimination or reduction of trade barriers between the United States and Mexico will accelerate the industrial integration that is already taking place between the two economies, which will mean more industrial activity along the border because of free trade. Incorporating environmental issues as part of trade negotiations was resisted by both governments, but U.S. congressional leaders have insisted that the two countries begin talks soon on the impact of further economic integration on the environment and shared natural resources.

A groundwater treaty could establish monitoring procedures so that the location and extent of pollution sources can be determined. It could specify locations permissible for the dumping of industrial waste and safety measures with which industries on both sides would have to comply. This would mean locating aquifer recharge zones and protecting these areas by passing appropriate laws. Finally, a treaty could empower state as well as federal agencies to enforce compliance with the standards delineated in the treaty. A groundwater treaty might also regulate the amount of groundwater withdrawals and control the kinds and quantity of residuals, such as pesticides and nitrates, that are carried to aquifer recharge zones. This would involve harmonizing standards for agricultural chemicals so that pesticides illegal in one country are also prohibited in the other.

A Recommended Framework

The groundwater supply and quality problems within the Rio Grande/Río Bravo Basin are so severe that the option of continuing with no change in the status quo should be excluded. Allowing each side to try to out-pump the other creates a climate of uncertainty regarding scarce water resources and increases the risks of irreversible damage to regional aquifers. The issue is not whether Mexico and the United States will someday regulate groundwater separately or together, but how they can adopt procedures that protect respective national interests while supporting the aspirations of their citizens to a reliable and sustainable supply of quality groundwater.

It is likely that any Mexican-U.S. groundwater treaty would begin by expanding the authority of the IBWC to regulate groundwater. Since 1944 both nations have jointly stated on many occasions that the IBWC has jurisdiction over border water allocation and sanitation issues. Thus two nations could authorize an extension of the IBWC's jurisdiciton to include eight new tasks associated with coordinated management of groundwater:

1. Commission research into geohydrology.
2. Identify the location, volume, and quality of current groundwater, by aquifer.

3. Collect data to establish the rate of groundwater withdrawal and recharge by aquifer.
4. Monitor surface water discharges that could affect groundwater quality.
5. Model groundwater migration, recharge, and depletion under various scenarios.
6. Assess how each side uses groundwater resources and its role in the local economies.
7. Identify any potential capital investments which could be constructed cooperatively to enhance the sustained yield or quality of groundwater.
8. Resolve any disputes after the adoption of any bilateral framework through the Minute Order process established in the 1944 Treaty.

To accomplish these goals expeditiously, the IBWC/CILA could be authorized to involve groundwater experts from federal agencies (such as EPA and Sedesol), affected state or local agencies, water users, and the engineering and academic communities in binational advisory groups. Some of the issues that such advisors might address are: (a) means for regulating firms or individuals who seek to discharge wastes below a shared international aquifer and (b) implementing regulations on the proper application of pesticides and nutrients in aquifer recharge zones.

A system that drives everyone to pump and divert as much water as possible is a fast way to drain the aquifers. This present laissez faire groundwater regulatory system is of no long-term benefit to either nation or current water users, even if it is expedient. New and existing economic investments will be jeopardized if chronic water shortages or pollution undermine the economic viability of border twin cities. It would be ironic if the United States and Mexico, two instrumental leaders in the multilateral negotiation to phase out ozone-depleting chemicals, could not address the sustainable supply and quality of the waters under their shared border.

Conclusion

Continuing to manage groundwater independently on each side of the border without any bilateral agreement, an expedient option in the short run, is of no long-term benefit to either nation or current water users. New and existing investments will be jeopardized if chronic water shortages destroy the economic viability of the border's urban areas. Absent an enforceable agreement for equitable allocation, communities will be driven to pump and divert as much water as they can before their sister cities across the border do the same, a fast way to drain the aquifers. It would be truly ironic if the United States and Mexico, both of whom were instrumental leaders in multilateral negotiations to phase out ozone-depleting chemicals, could not come to any bilateral agreement on an environmental challenge at their own doorstep.

Notes

1. Texas Water Commission, *Ground-Water Quality of Texas,* Austin, Texas , 1989, pp. 7-16.

2. U.S. Geological Survey, cited in Jeffrey L. Strause, *Texas Ground-Water Quality,* OFR 87-754, p. 1.

3. Charles Webster, Texas Water Commission, Weslaco field office, telephone conversation with David Hurlbut, July 7, 1992.

4. TWC, *Groundwater Quality,* p. 155.

5. TWC, *The State of Texas Water Quality Inventory, 9th Edition,* April 1988, p. 563.

6. Randall J. Charbeneau, "Groundwater Resources of the Texas Rio Grande Basin," *Natural Resources Journal,* Vol. 22, No. 4, October 1982, p. 959.

7. Neal E. Armstrong, "Anticipating Transboundary Water Needs and Issues in the United States-Mexico Border Region," *Natural Resources Journal,* Vol. 22, No. 4 , October 1982, p. 896.

8. TWC, *Groundwater Quality,* p. 171.

9. María del Rosario Díaz A. and Alfredo Granados Olívas, "Evaluacion Geohidrológica del Acuifero de la Zona Urbana de Ciudad Juárez, Chihuahua, Período 1980-1990," unpublished paper, Instituto de Ingenieria y Arquitectura de la Universidad Autónoma de Ciudad Juárez, September 1991.

10. TWC, *Groundwater Quality*, p. 134.

11. International Boundary and Water Commission, "Flow of the Rio Grande and Related Data," Water Bulletin No. 60, 1990. See table for "Diversions from the Rio Grande, American Canal at El Paso, Texas."

12. Beverly Ray, Texas Department of Health Epidemiology Division, telephone interview with David Hurlbut, July 29, 1992.

13. Charbeneau, "Groundwater Resources," pp. 959-960.

14. TWC, *Water Quality Inventory*, p. 563.

15. TWC, *Water Quality Inventory*, p. 959.

16. Texas Department of Water Resources, *Occurrence and Quality of Ground Water in the Edwards-Trinity (Plateau) Aquifer in the Trans-Pecos Region of Texas,* Report 255, September 1980, pp. 11-15.

17. Richard Ginn, Texas Railroad Commission, telephone conversation with David Hurlbut, July 7, 1992.

18. Charles Webster conversation with David Hurlbut, July 7, 1992.

19. Jeffrey L. Strause, *Texas Groundwater Quality*, p. 1.

20. International Law Association, *Helsinki Rules on the Uses of the Waters of International Rivers,* London, 1967, cited in Robert D. Hayton and Albert E. Utton, "Transboundary Groundwaters: The Bellagio Draft Treaty," *Natural Resources Journal,* Vol. 29, Summer 1989, p. 669.

21. United Nations Economic Commission for Africa, and Department of Technical Co-operation for Development, "Interregional Meeting on River and Lake Basin Development, with Emphasis on the Africa Region," Addis Ababa, Ethiopia, October 10-16, 1988; cited in Robert D. Hayton and Albert Utton, "Transboundary Groundwaters," p. 671.

22. Robert D. Hayton and Albert Utton, "Transboundary Groundwaters," p. 672.

23. Stephen P. Mumme, "The United States-Mexico Groundwater Dispute: Domestic Influence on Foreign Policy," Ph.D. dissertation, University of Arizona, December 1981, p. 331.

24. C. Richard Bath, and Dilmus D. James, "Transborder Flows of Technical Information: Cases of the Commercialization of Guayle and Groundwater Utilization," Working Paper, U.S.-Mexico Project Series, No. 10, Overseas Development Council, July 1982, p. 13.

25. Robert D. Hayton and Albert E. Utton, "Transboundary Groundwaters," pp. 663-722.

26. "Arrangement relatif a la protection, a l'utilisation et la realimentation de la nappe souterraine franco-suisse du genevois," groundwater treaty between France and Switzerland.

27. Stephen P. Mumme, "The United States-Mexico Groundwater Dispute," p. 327.

28. See Articles 5, 6, 9, and 11 of the 1987 Montreal Protocol on Substances that Deplete the Ozone, and Article 10 of the 1990 London Revisions to the Montreal Protocol.

29. Stephen P. Mumme, "The United States-Mexico Groundwater Dispute," p. 327.

30. Position expressed by a Texas court in the case of *City of Corpus Christi v. City of Pleasanton*. As quoted in "Texas Ground Water Law: A Survey and Some Proposals," Corwin W. Johnson, p. 2.

31. See Texas Department of Water Resources, *Water for Texas: Technical Appendix*, November 1984, Volume 2, p. I-16.

32. TDWR, *Water for Texas*, p. I-19.

33. As quoted in Stephen P. Mumme, "The United States-Mexico Groundwater Dispute," p. 181.

34. Mexico's Social Welfare Secretariat (Sedesol) has some role in matters concerning groundwater conservation. The Programming and Budget Secretariat (SPP), now part of the Treasury Secretariat, and the Federal Electricity Commission (CFE) are playing a more significant role in affairs related to the budget and the oversight of policy. See Stephen P. Mumme, "The United States-Mexico Groundwater Dispute," pp. 184-85.

35. As quoted in Stephen P. Mumme, "The United States-Mexico Groundwater Dispute," p. 297.

Chapter 4:
Water Quality Conflicts

WATER QUALITY IN THE RIO GRANDE/RÍO BRAVO IS BAD and getting worse. Much of the problem is the untreated wastes that are piped into the river from the twin cities along the border. The water in the river has many contaminants and bacterial levels that violate both Texas and Mexican standards. Although the U.S. side is not blameless, the majority of the untreated wastes comes from the Mexican side. Poor water quality of the Rio Grande/Río Bravo undoubtedly contributes to relatively high rates of water-borne disease along the border.

One solution to these water quality problems would be the collection of wastewaters within Mexican cities and subsequent secondary sewage treatment. Numerical models that assume historical point-source effluent emissions and concentrations from Nuevo Laredo indicate that conventional secondary treatment can produce effluent which, when mixed with the waters of the Rio Grande/Río Bravo, yield a water quality that meets both U.S. and Mexican standards. Institutional mechanisms now exist for Mexico and the United States to cooperate in joint wastewater collection and treatment projects.

Although wastewater treatment would improve surface water quality in the Rio Grande/Río Bravo, a number of problems will continue to exist. These include the contributions of non-point sources of water pollution, releases of hazardous or toxic wastes into the water, the inadequate level of binational funding for waste collection and treatment, and the post-construction performance of the treatment systems.

This chapter describes the current quality of the water in the Rio Grande/Río Bravo and its consequences and discusses the technical solutions that are available and their likely effects. The chapter also addresses remaining uncertainties that will exist regarding the future conflicts between Mexico and the United States over the quality of their shared border river.

Quality of the Rio Grande/Río Bravo

Water quality within the Rio Grande/Río Bravo Basin has been declining for the past 20 years as increasing volumes of untreated sewage are discharged into the river. In 1983, the U.S. Geological Survey reported on a long-term study of discharges at some 504 surface water quality monitoring stations throughout the United States. The survey reported on dissolved oxygen, perhaps the best indicator of overall surface water quality, for the period of 1974 to 1981. Any decrease in dissolved oxygen would indicate degradation of stream quality. In most areas of the nation, the number of stations showing an increase in stream water quality exceeded the number of stations reporting poorer water quality. This was not the case in the Rio Grande/Río Bravo Basin, where the number of stations with decreases (degradation) exceeded the number with increases (improvements).[1]

A site-by-site assessment based on the 1980 water year also indicates decreasing water quality within the Rio Grande/Río Bravo. Figures 4.1 and 4.2 illustrate the minimum, median, and maximum values of total dissolved solids and fecal coliform levels.[2] Total dissolved solids (TDS) is an indicator of water "sweetness" and potential for use either as a source for drinking water or for irrigation; the higher the TDS value, the lower the water quality. TDS was worst around Fort Quitman, Langtry, the Río Salado, and Morillo Drain; the levels violated Texas surface water quality standards. Fecal coliform is an indicator organism for organic pollution used to assess a river's fitness for use in water recreation or as a source for drinking water. Coliform concentrations exploded at sites below El Paso/Juárez, Laredo/Nuevo Laredo, and Brownsville, Matamoros to levels far much worse than Texas' surface water quality standards.

To determine whether water quality in the river had improved since these earlier studies, Rio Grande/Río Bravo fecal coliform measurements were evaluated for the period of August 1, 1986 through July 31, 1989.[3] To discuss the results, it will be useful to compare Texas and Mexican stream water quality standards. As indicated in Table 4.1, the Texas Water Commission (TWC) has established a numerical standard for bacterial contamination within all but one portion of the Rio Grande/Río Bravo: an average of no more than 200 coliform colonies per 100 milliliter sample.[4] The Mexican Secretaría de Agricultura y Recursos Hidráulicos (SARH) set two criteria for surface water such as the Río Bravo to be used as a source for drinking water after sand filtration and chlorination.[5] These are a monthly average that does not exceed 1,000 colonies per 100 milliliter sample, and no more than 10 percent of the samples with more than 2,000 colonies per 100 milliliter.[6] (See Table 4.1.) Both Texas and SARH set maximum upper limits for water used for non-contact recreation, such as boating, or by industry; the Mexican upper level is 10,000 colonies and the Texas upper limit is 2,000, each calculated per 100 milliliter sample.[7]

Table 4.1 indicates that the Rio Grande/Río Bravo violated Texas surface water quality standards frequently during the period of August 1, 1986 through July 31, 1989. Fecal coliform counts were usually in violation of standards below each of the twin cities except El Paso/Ciudad Juárez. The river on some occasions exceeded other surface water quality standards, including dissolved oxygen and chlorides below Presidio/Ojinaga, Del Rio/Acuña, Eagle Pass/Piedras Negras, Laredo/Nuevo Laredo, and Hidalgo/Reynosa.[8]

Table 4.2 provides more detailed information on fecal coliform levels above and below five twin metropolitan regions.[9] Average coliform levels above each of the twin cities did not exceed either Texas or Mexican surface water quality standards. Average coliform levels below each region exceeded both the Texas and Mexican criteria. The coliform levels below the cities of Laredo and Nuevo Laredo were 75 times higher than the Texas criteria for drinking water. Water below Laredo/Nuevo Laredo had coliform levels far above Texas and Mexican standards for industrial and non-contact recreation.

More recent surface water quality measurements confirm fecal coliform contamination remains endemic at Laredo/Nuevo Laredo. The Austin American-Statesman reported levels exceeding 1,000 times Texas' upper limit for water recreation of 200 colonies per 100 milliliter.[10] Individual coliform tests have recorded as high as 53 million colonies per 100 milliliter sample below Laredo/Nuevo Laredo.[11]

Coliform bacteria indicate pollution, but they do not communicate the full range of the industrial wastes, pesticides, and other hazardous and/or toxic materials added by the 3.5 mil-

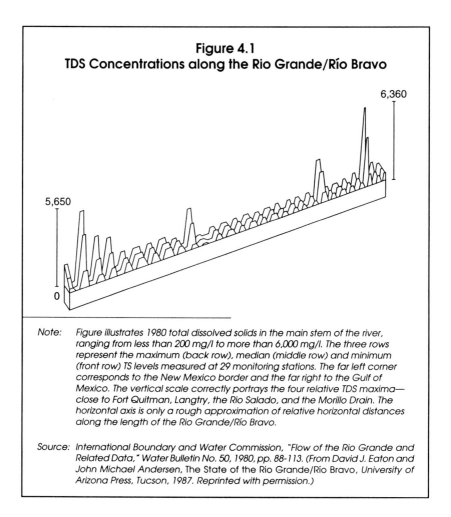

Figure 4.1
TDS Concentrations along the Rio Grande/Río Bravo

6,360

5,650

0

Note: Figure illustrates 1980 total dissolved solids in the main stem of the river,
          ranging from less than 200 mg/l to more than 6,000 mg/l. The three rows
          represent the maximum (back row), median (middle row) and minimum
          (front row) TS levels measured at 29 monitoring stations. The far left corner
          corresponds to the New Mexico border and the far right to the Gulf of
          Mexico. The vertical scale correctly portrays the four relative TDS maxima—
          close to Fort Quitman, Langtry, the Rio Salado, and the Morillo Drain. The
          horizontal axis is only a rough approximation of relative horizontal distances
          along the length of the Rio Grande/Río Bravo.

Source: International Boundary and Water Commission, "Flow of the Rio Grande and
          Related Data," Water Bulletin No. 50, 1980, pp. 88-113. (From David J. Eaton and
          John Michael Andersen, The State of the Rio Grande/Río Bravo, University of
          Arizona Press, Tucson, 1987. Reprinted with permission.)

lion people whose communities are drained by the Rio Grande.
Neither Texas nor Mexico has undertaken a complete water
quality survey to test for the range of heavy metals or organic
chemicals within the river.

What are the consequences of foul surface water in the Rio
Grande/Río Bravo? Evidence indicates that human health has
been affected, particularly in Mexico where fewer people have
access to properly treated drinking water. One study comparing Mexican and Texas rates of water related illness found Mexican amebiasis rates 500 times as high as rates in Texas in 1978.[12]
The same study found mortality due to enteric diseases to be

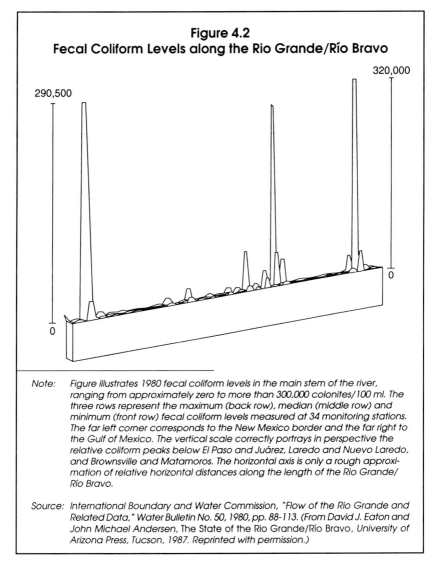

Figure 4.2
Fecal Coliform Levels along the Rio Grande/Río Bravo

Note: Figure illustrates 1980 fecal coliform levels in the main stem of the river, ranging from approximately zero to more than 300,000 colonites/100 ml. The three rows represent the maximum (back row), median (middle row) and minimum (front row) fecal coliform levels measured at 34 monitoring stations. The far left corner corresponds to the New Mexico border and the far right to the Gulf of Mexico. The vertical scale correctly portrays in perspective the relative coliform peaks below El Paso and Juárez, Laredo and Nuevo Laredo, and Brownsville and Matamoros. The horizontal axis is only a rough approximation of relative horizontal distances along the length of the Rio Grande/Río Bravo.

Source: International Boundary and Water Commission, "Flow of the Rio Grande and Related Data," Water Bulletin No. 50, 1980, pp. 88-113. (From David J. Eaton and John Michael Andersen, The State of the Rio Grande/Río Bravo, University of Arizona Press, Tucson, 1987. Reprinted with permission.)

significantly higher along the Mexican side of the river, as compared to adjacent Texas counties.[13] A second study found that post-neonatal mortality (deaths occurring after 30 days of life) were seven times more common in Nuevo Laredo than in Laredo.[14] The study also found that "bronchitis and pneumonia, noninfectious gastroenteritis, and intestinal infections accounted for 184 deaths in infants in the post-neonatal period in Nuevo Laredo, while only two children in Laredo died of these causes."[15]

A causal connection can be made between polluted surface water and human health. In 1980, for example, El Paso County reported 37.5 infectious hepatitis cases per 100,000 population, compared to a U.S. rate of 12.8, and Webb County reported salmonellosis of 55.4 cases per 100,000, versus the U.S. rate of 14.9 cases.[16] The Texas Attorney General submitted the following testimony to the EPA in 1991 about life in colonias where water and wastewater systems are inadequate.

> People will haul water and store it in 55-gallon drums. The El Paso County Health Department conducted a survey of one colonia that had no potable water and found that about half the people stored water in this type of drum and that 70 percent of the drums were labeled indicating the contents were toxic, such as methylene chloride, stoddard solvent, and trichlomethane.
>
> The health consequences of these living conditions are enormous. The University of Texas Health Science Center in San Antonio conducted a study of certain colonias and found that by the age of 8 years approximately 55 percent of the colonia children had already been infected with hepatitis A and by the age of 35 years, 85 percent to 99 percent of the residents had been so infected. In El Paso County, the rate of hepatitis, tuberculosis, and dysentery are three to five times higher than the national average. Figures for the Rio Grande Valley are worse.[17]

The source of the pollution has been known for many years: millions of gallons of untreated or partially treated sewage and industrial waste flows from the twin cities along the Rio Grande/Río Bravo every day. On the Texas side, much progress has been made to treat sewage prior to release to the river so that discharges meet state emissions standards. Much less progress has occurred on the Mexican side.

Table 4.3 lists the pattern of discharges of Texas wastewater treatment plants in three of the twin metropolitan areas. The TWC discharge standards for each of the plants represent con-

Table 4.1
Water Quality Problem Areas along the Rio Grande/Río Bravo

Twin cities	Segment	Surface water standard (coliform colonies/100 ml) Texas	Mexico	Monitoring stations	Violation of Texas Standards coliform (%)	other (%)
El Paso/ Juárez	2314	200-avg.	1000-avg. 2000-max.	.0100 .0700	32 18	-
	2308	2000-avg.	1000-avg. 2000-max.	.0500	none	-
Presidio/ Ojinaga	2306	200-avg.	1000-avg. 2000-max.	.0300	39	chloride: 45
Del Rio-Acuña	2304	200-avg.	1000-avg. 2000-max.	.0250 .0300	58 13	pH: 6
Eagle Pass/ Piedras Negras	2304	200-avg.	1000-avg. 2000-max.	.0150 .0200	81 23	pH: 11 pH: 14
Laredo/ Nuevo Laredo	2304	200-avg.	1000-avg. 2000-max.	.0050 .0075 .0090 .0095	94 100 100 30	chloride: 25
Hidalgo/ Reynosa	2302	200-avg.	1000-avg. 2000-max.	.0150 .0200 .0250 .0300	62 50 n/a n/a	DO: 6 DO: 11
Brownsville/ Matamoros	2302	200-avg.	1000-avg. 2000-max.	.0100	55	-

Note: The pH standard is that the acid/base balance is between 6.0 and 9.0. The DO or dissolved oxygen standard is a minimum of 5.0 mg/l. The numbers in the right hand column are violations as a percent of all tests.

Source: Compiled by Andrew Sherrill from unpublished Texas Water Commission records for the three-year period August 1, 1986 to July 31, 1989.

ventional performance for secondary waste treatment, maximum emissions of 20 milligrams per liter (mg/l) biochemical oxygen demand and 20 mg/l total suspended solids. All of the plants meet or exceed these standards.[18] All but the waste treatment plants from Del Rio discharge lower volumes of waste than authorized by the TWC. Altogether these four areas contribute on the order of 20 million gallons a day of sewage treated through conventional secondary processes.

A number of factors contribute to the volume of untreated wastewater effluent discharged from the Mexican side of the Rio

Table 4.2
Fecal Coliform Levels Above and Below
Twin-City Discharges, 1986-1989
(number of colonies per 100 milliliter of water)

Monitoring station (location from cities)	Quality criterion	Maximum	Average	Number of samples	Number above criterion
El Paso/Juárez					
2304.0300 (upstream)	200	480	81	33	4
2304.0250 (downstream)	200	6,800	692	33	19
2304.0350 (city creek)	200	3,800	370	28	9
Eagle Pass/Piedras Negras					
2304.0200 (upstream)	200	410	143	31	7
2304.0150 (downstream)	200	3,332	667	31	25
Laredo/Nuevo Laredo					
2304.0100 (upstream)	200	580	75	32	2
2304.0095 (in city)	200	700	190	10	3
2304.0090 (in city)	200	4,700	2,694	7	7
2304.0050 (80 miles downstream)	200	30,000	6,051	34	32
Hidalgo/Reynosa					
2304.0200 (in city)	200	5,200	565	32	16
2304.0150 (in city)	200	22,500	2,013	29	18

Note: Water below Eagle Pass/Piedras Negras also violates the pH criteria set by
the Texas Water Commission, a maximum level of pH 9.0. Five of 36 tests at
station 2304.0200 and four of 37 tests at station 2304.0150 exceeded pH of
9.0.

Source: An analysis by Andrew Sherrill of water quality data for the period August 1,
1986, to July 31, 1989, from unpublished computer printouts furnished by the
Texas Water Commission's statewide monitoring network, Austin, Texas,
December 1, 1989.

Grande/Río Bravo. Not all of the wastewater is treated in the
Mexican *municipios* (Ciudad Juárez, Ojinaga, Ciudad Acuña,
Piedras Negras, Nuevo Laredo, Ciudad Mier, Camargo, Reynosa,
or Matamoros) due to the limited funds to construct collection
systems. A substantial portion of the collected wastewater cannot be conducted to treatment, because many of the clay and
concrete pipes that exist are broken, collapsed, or they leak.[19]
Effective wastewater treatment plants have not been built ev-
erywhere that sewerage systems exist, and those that do exist
do not always operate properly.

As a consequence, more than 60 million gallons a day of untreated or partially treated sewage and industrial waste flow

Table 4.3
Discharges from the U.S. into the Rio Grande, 1989

Discharger	Flow level (millions of gal./day) Permit	Actual	BOD$_5$ level (milligrams per liter) Permit	Actual	Total suspended solids (milligrams per liter) Permit	Actual
City of Del Rio						
Silver Lake Plant	1.76	1.80	20	8.2	20	9.2
San Felipe Plant	1.63	1.73	20	3.6	20	8.7
Round Mtn. Plant	0.61	0.82	20	6.5	20	11.8
City of Eagle Pass	3	2.6	20	16	20	18
City of Laredo						
Zacate Creek Plant	14	9.1	20	13	20	20
Southside Plant	3	2.4	20	5	20	12
Starr County CID	1.5	0.5	20	13	20	6
City of Rome	0.36	0.34	20	18	20	11.5

Note: BOD$_5$ is biological oxygen demand of wastes, measured at day 5. The values are the maximum monthly levels reported in the first nine months of 1989.

Source: Texas Water Commission, "Self-Reporting Wastewater Discharge Data for the Rio Grande Basin," Austin, Texas, December 1, 1989 (computer printout).

from five of the Mexican cities (Ciudad Juárez, Ciudad Acuña, Piedras Negras, Nuevo Laredo, and Reynosa) into the Rio Grande. Table 4.4 contains available information on waste discharge from the Mexican side of the Rio Grande/Río Bravo. The problem in Juárez is the lack of wastewater treatment capacity. Acuña has two sewage by-passes around a sewage lagoon, again reflecting inadequate capacity. Nuevo Laredo has an incomplete wastewater collection system and no operating sewage treatment, although one is under construction. Reynosa has treatment lagoons but the discharges still show high coliform counts indicating inadequate treatment.[20]

John Hall, Chairman of the Texas Water Commission, has described the problem and one solution:

> Water quality standards are being violated downstream of virtually every pair of sister cities on the river. Wastewater treatment plants are needed for Juárez, Reynosa, Matamoros, and other cities. This problem is obvious and merits immediate action.[21]

Table 4.4
Some Discharge Sites along the
Mexican Side of the Rio Grande/Río Bravo

Location	Volume (millions of gallons per day)	Contents	Site of quality measurement	Coliform count (colonies/100ml)
Ciudad Juárez	30	sewage, industrial waste	n/a	n/a
Ciudad Acuña	2.5	sewage	2300.200	16,000 (1988) 110,000 (1988)
	0.2	sewage	2300.9002	1.4 million (Mar. 1988) 26 million (June 1988)
Piedras Negras	2.5	sewage	2300.1700	85,000 (Mar. 1988) 19,000 (June 1988)
Nuevo Laredo	20	sewage	28 drains	95% above 60,000
	5	sewage	non-point	max.: 53 million
Reynosa	n/a	sewage	n/a	n/a

Sources: James Garcia, "Border river laden with wastes," Austin American Statesman, September 29, 1991, p. A17, Col. 2; Narendra N. Gunaji, "Memorandum to the Texas Water Commission: Status of Border Sanitation Problems as of December 12, 1988," El Paso, Texas, December 14, 1988; David L. Buzan, "Intensive Survey of the Rio Grande: Segment 2304," draft report, Texas Water Commission, Austin, Texas, June 1989, Table II.

Who is Responsible and Who Will Pay?

If wastewater collection and treatment is a solution to the water quality problems of the Rio Grande, three related questions are: (a) who will treat the wastes; (b) what treatment is appropriate; and (c) how will water quality change after wastewater is collected and treated?

Under international law, Mexico is responsible for the continuing degradation because the wastewater is being discharged from Mexican communities and affecting the rights of U.S. communities to use the Rio Grande. However there are mitigating factors, including the role of U.S.-oriented maquiladora industries in generating wastes and the shared jurisdiction implied by the Treaty of 1944 and subsequent bilateral agreements. As a practical matter, mechanisms exist for the two nations to cooperate in building sewage collection and treatment systems to

Table 4.5
International Judicial Decisions
Affecting Rio Grande Water Quality

Case	Tribunal holding
Trail Smelter	"Under the principles of international law ... no state has the right to use or permit the use of its territory in such a manner as to cause injury ... in or to the territory of another ... when the case is of serious consequence and the injury is established by clear and convincing evidence."
Lake Lanoux	A state has an obligation to consult and negotiate if it plans to take actions that could affect adversely the interests of another state.
Corfu Channel	Every state has an obligation not to allow knowingly its territory to be used for acts contrary to the rights of other states.
Gasser v. U.S.	The United States is obligated to protect property in Mexico from harm caused by the operations of U.S. firms.

Sources: for Trail Smelter and Lake Lanoux, Albert Utton, "International Water Quality Law," in International Environmental Law, edited by Teclaff and Utton,1975, p. 154; for Corfu Channel, Nisuke Ando, "The Law of Pollution Prevention in International Rivers and Lakes, in The Legal Regime of International Rivers and Lakes, edited by Zacklin and Caflish, 1981, p. 333; for Gasser v. U.S., Morgan, Anne M., "Transboundary Liability Goes with the Flow? Gasser v. United States: The Use and Misuse of a Treaty," in Transboundary Resources Report, Vol. 5 , No. 12, (Summer 1991), p. 3.

be operated by the Mexican side under binational supervision. Federal, state, and local governments on the U.S. side can contribute funds to construct and operate the wastewater infrastructure.

Mexico has a duty under international law to reduce the pollution of the Rio Grande/Río Bravo because the discharges from its side preclude some American beneficial uses of the river. There are two lines of argument, one from case law and the other from international principles.[22] Four cases establish a state's responsibility for its pollution damaging citizens of another state (see Table 4.5) and three sets of international legal agree-

Table 4.6
International Law Resolutions Affecting Rio Grande Water Quality

Source	Document	Rules
Int'l. Law Association	Helsinki Rules, Article IV	"Each basin state is entitled within its territory to a reasonable and equitable share in the beneficial uses of the waters of an international drainage basin."
Int'l. Law Association	Helsinki Rules, Article X	"Consistent with the principle of equitable utilization" a state "must prevent any . . . increase in the degree of existing water pollution in an international drainage basin which would cause substantial injury in the territory of a co-basin state, and should take all reasonable measures to abate existing water pollution."
UN Conference on the Human Environment	Stockholm Principles, Principle 21	"States have . . . the responsibility to ensure that activities within their jurisdiction or control do not cause damage to the environment of other states."

Sources: International Law Association, Helsinki Rules, 1966, Article IV; Ando, p. 345; Report of the United Nations Conference on the Human Environment.

ments confirm that sovereignty does not protect Mexico from an obligation to abate its sewage discharges (see Table 4.6).

The cases of Trail Smelter, Lake Lanoux, Corfu Channel, and Gasser set international precedents for national responsibility for actions that damage another state or its citizens. Corfu Channel[23] and Trail Smelter[24] establish that Nation A can be held liable for injury to persons in Nation B caused by acts in A's territory, if damages are serious and supported by convincing evidence. Lake Lanoux[25] holds that Nation A must consult with Nation B if A's action may cause injury in B. Gasser[26] has been interpreted to rule that Nation A is obligated to protect B's citizens from harm if the injury is caused by A's infrastructure. In other words, Mexico has an obligation to consult with the United States about the discharge of sewage, abate those emissions, and protect U.S. citizens from injury caused by that pollution if the damages are serious and supported by convincing evidence.

On three occasions groups of international lawyers have gathered to develop principles for management of international rivers. Table 4.6 lists the rules promulgated by the International Law Association, the United Nations Conference on the Human Environment, and the International Law Commission.[27] Their legal principles develop the theme of "limited sovereignty." A nation in an international river basin may make beneficial uses of the waterway as long as its actions do not damage the rights of others or preclude other riparian nations from beneficial uses of the same river.

Mexico is the origin of the pollution which precludes two U.S. uses of the Rio Grande (contact and non-contact recreation) and makes more expensive and increases the health risks of a third use (drinking water). Mexico is responsible under the principles of international courts and international law for abatement of that pollution. Mexican per capita income in border areas is between one-fifth and one-seventh of the per capita incomes of U.S. border states, and Mexico is dependent on the United States in many ways.[28] Nevertheless, the effluent from its cities causes conditions in gross violation of Mexico's own surface water quality standards.

While the United States might win a case in international courts against Mexico, the issue is not likely to be brought to such a tribunal, as the two states have a treaty and other international agreements authorizing cooperative work reducing pollution. The U.S.-Mexican Water Treaty of 1944 established a binational commission to deal with mutual border sanitation problem. In the United States it is known as the International Boundary and Water Commission (IBWC) and in Mexico as the Comisión Internacional de Límites y Aguas (CILA). The Treaty gives the commission jurisdiction over:

> any sanitary measures or works which may be mutually agreed upon by the two Governments, which hereby agree to give preferential attention to the solution of all border sanitation problems.[29]

The IBWC/CILA adopted Minute Order No. 261 in September 1979 that interprets the clause "all border sanitation problems" to include all classes of water pollution, subject to a specific agreement that must be signed committing the Commission to act in any particular case.[30] This minute order has been implemented in two subsequent minute orders, No. 264, relating to Calexico, California/Mexicali, Baja California, and No. 279 for Laredo/Nuevo Laredo.[31] The U.S. Congress has indicated a willingness to consider additional joint projects through the Rio Grande Pollution Correction Act of 1987.[32] This law authorizes correction of international pollution problems caused by discharge of inadequately treated sewage from Texas/Mexico border cities, including Ciudad Acuña, Nuevo Laredo, Reynosa, Del Rio, Laredo, and Hidalgo.

Mexico and the United States signed an Executive Agreement on August 14, 1988 expanding cooperation on environmental problems through coordinated planning by the EPA and the Mexican Secretaría de Desarrollo Urbano y Ecología (now Secretaría de Desarrollo Social, or Sedesol).[33] This Agreement:

- established EPA/SEDUE cooperation to monitor environmental issues;
- authorized protection of the border environment through conclusion of subsequent specific agreements (Annexes) and coordinating domestic programs, as well as exchange of information, data and staff; and
- provided for state, local, and non-governmental organization involvement in environmental deliberations.

This bilateral cooperation has been implemented through a joint environmental planning document for the border region.[34]

The cooperative planning and construction of sewage collection and treatment is a logical solution given the origins of the wastes: the pollution derives from a border economy and not just Mexican *municipios*. More than 260 foreign-owned factories, called maquiladoras, operate around Juárez/El Paso and there are over 100 maquiladoras around Nuevo Laredo/Laredo.[35] Nearly 2,000 maquiladoras on the U.S.-Mexico border (40 percent of which are along the Rio Grande) shipped $12.8 billion in exports to the United States in 1990, of which half consisted of products shipped orginally to Mexico from the United States and half value added in Mexico.[36]

Cooperative planning and construction implies joint funding. The Treaty of 1944 is silent on the split of funds between Mexico and the United States for water quality improvement,[37] although the general principle is to determine the cost allocation on a case-by-case basis. It says:

> The cost of construction, operation, and maintenance of each of the . . . joint works . . . shall be prorated between the two Governments in proportion to the benefits . . . as determined by the Commission and approved by the two Governments.[38]

In the case of the Laredo/Nuevo Laredo wastewater collection and treatment project, the Commission decided to divide the estimated $35 million in costs equally between the United States and Mexico.[39] The State of Texas has contributed to the funding.

Although the Rio Grande Pollution Control Act authorizes the IBWC/CILA to consider construction of wastewater collection and treatment for twin cities along the border, funds have been authorized only for the Laredo/Nuevo Laredo project. The recent joint environmental plan is silent on a timetable to improve water quality elsewhere along the river.[40] It remains to be determined when or where additional pollution control infrastructure will be constructed.

In summary, even though the discharges of untreated wastewater are of Mexican origin, the two nations will be planning, constructing, operating, and financing treatment infrastructure together.

What Will Be Built and Will It Work?

The water quality problems of the Rio Grande—excessive bacterial contamination and insufficient dissolved oxygen—reflect inadequately treated domestic and industrial sewage. The section below describes the range of choice for Mexico and the United States for wastewater collection and treatment and evaluates whether conventional solutions can produce discharges that will meet Mexican and Texas standards.

Water pollution enters a stream either via a point source (such as a discharge pipe or ditch) or through diffuse runoff from the land (a non-point source). In the United States, the conventional approach is to collect sewage from homes and factories in the same urban drainage pipes that carry storm water, a combined sewer system. Under conditions of normal wastewater flow, sewage is conveyed through a collection system by gravity or pumping and delivered to a wastewater treatment plant that has been built to handle a specified flow of wastes. With heavy rains, only a portion of the combined sewage and storm flow will be treated: flow above the capacity of the treatment plant will be discharged directly in to the stream. Hazardous wastes or toxic materials should be prohibited in the sewer system. Such toxics should be pre-treated by the factory that emits them or disposed in some other manner.

Primary treatment in most sewage treatment processes consists of removal of large objects (screening) and the use of gravity to settle large particles (grit chamber) and smaller particles and organic solids (settling tank). Primary treatment can remove about 35 percent of the organic materials from typical municipal sewage.

Secondary treatment relies upon one of three biological processes: activated sludge, a trickling filter, or oxidation lagoons/ditches. Each allows bacteria in wastewater sufficient time and the proper conditions to break down sewage through their growth. A trickling filter is a bed of stones that provides microorganisms with places to grow as the sewage trickles down via gravity. Activated sludge takes place in a tank that mixes air and sewage with microorganisms to yield a concentrated mass of growth (sludge) along with carbon dioxide and water. The wastewater is passed through a second settling tank to remove sludge. About 90 percent of the organic matter of sewage can be removed through primary and secondary treatment. Sewage lagoons or ditches take the wastewater after primary treatment into a lagoon or ditch. Natural biological processes allow bacteria to live off wastewater nutrients, reducing organic wastes in the process. Aerated lagoons or oxidation ditches can reduce organics to levels that meet U.S. federal and state standards.

The term tertiary treatment is used to refer to a variety of specialized chemical processes designed to remove nitrogen and phosphorous, which river plants use as nutrients. Excessive nutrients can fuel explosive plant growth and the reduction of dissolved oxygen, leading to eutrophication, the modification of the stream ecosystem. Tertiary treatment may be followed

Table 4.7
Water QualityStandards for the Rio Grande

Indicator	Standard
Dissolved oxygen	Not less than 5.0 milligrams per liter[1]
pH	Between 6.5 and 9.0
Fecal coliform	200 colonies per 100 milliliter sample[2]
Other indicators[3]	varies by river segment

[1] Except for Devils River to International Dam (6.0 mg/l), International Dam to Riverside Diversion Dam (3.0 mg/l), and segment immediately below Riverside Dam (3.0 mg/l when flow over the dam is less than 35 cubic feet per second).

[2] Except for segment from International Dam to Riverside Diversion Dam (2,000 colonies per 100 ml).

[3] Total dissolved solids, chloride content, and sulfate content .

Source: 31 Texas Administrative Code Sec. 307, Appendix A.

by additional settling to remove solids. The sludge residues from primary, secondary, and tertiary treatment settling tanks must be disposed of, either through land application for the sludge or dried residue or further treatment (digestion) and disposal. Chlorine is added to kill bacteria before the treated wastewater is discharged into the environment.

The United States has adopted primary and secondary treatment as a minimum expectation of municipal wastewater treatment. Texas law requires that effluent standards for each treatment plant be computed on a case-by-case basis. These computations begin with overall water quality criteria that apply to an entire river segment (see Table 4.7), and take into account the volume of river flow, dispersion potential, current use, current quality of aquatic life, and the volume of discharge expected from the treatment plant.[41]

The Laredo/Nuevo Laredo Project has been designed to achieve at least the level of performance of Texas standards.[42] The project includes: two large pipes to collect existing wastewater; new sewage lines and renovations to existing pipe systems; a pumping plant; and secondary treatment using oxidation ditches, settling basins, and chlorination.[43] The design capacity of the system is 31 million gallons a day, sufficient for a Nuevo Laredo population of 600,000.[44]

Will such a wastewater treatment system reduce the pollution of the Rio Grande to levels that will meet Texas and Mexican standards for surface water to be used for drinking water, contact and non-contact recreation? One way to evaluate treatment plant performance in advance of construction is to simulate how treatment may reduce pollution. This was done by a research group at the Lyndon B. Johnson School of Public Affairs in 1990.

The first step was to obtain information regarding the amounts and locations of existing wastewater discharges in the river and the quality of the river both above and below the point source of pollution. Project participants used information from an intensive stream survey in March and June 1988 as the current information of Rio Grande surface water quality, the points of wastewater discharge along both Texas and Mexican sides, and the volumes of wastewater.[45]

The second step was to find a simulation model that has sufficient capacity, sophistication, and detail to model the treatment and river processes with reasonable validity. Project members selected the Qual-TX simulation developed by the state of Texas to evaluate other stream segments. Qual-TX can model a large number of stream segments reflecting numerous points of wastewater discharge of varying qualities, multiple water criteria, and a range of realistic stream flow conditions. Results from Qual-TX have been used by the Texas Water Commission to compute discharge standards for wastewater permit holders.

The third step was to simulate current conditions: known volumes of discharges from known points into known river flows. This simulation yielded conditions similar to those observed in the field, as determined through the intensive survey.

The fourth step was to model the effect of wastewater treatment on the quality of the Rio Grande/Río Bravo. Ziming Yang, a project member, simulated the surface water quality under two water flow regimes (March and June 1988) and five alternative levels of treatment:

- no treatment (existing condition);
- primary treatment of existing wastes;
- primary and secondary treatment of existing wastes;
- primary and secondary treatment plus chlorination; and

  • the proposed joint Laredo/Nuevo Laredo Project.

The simulation for primary treatment assumed that it would remove 35 percent of BOD_5, 35 percent of local suspended solids, and 50 percent of fecal coliform bacteria. The joint project run assumed collection of existing point discharges and flows unconnected to a sewer system into the river and treatment through the proposed secondary treatment plant, so as to produce the following effluents as designated by the IBWC/CILA principal engineers: 20 milligrams of BOD_5 per liter of water, 20 milligrams per liter total suspended solids, a dissolved oxygen concentration minimum of 2.0 milligrams per liter, and no more than 200 colonies of fecal coliform per 100 milliliters.[46]

Table 4.8 compares the quality of the Rio Grande, as simulated under no treatment, ordinary treatment, and joint project for June 1988 flow conditions with discharges as recorded in June 1988.[47] The hypothesis to be tested was whether the joint project would discharge wastes that allow the Rio Grande to meet stream standards for use as a source of drinking water, contact recreation, and non-contact recreation.

Although the simulation performed by Ziming Yang is not the final word on this subject, it does show that the collection and conventional secondary treatment of wastewater along both sides of the Rio Grande can improve the quality of the river up to Texas and Mexican standards. Some additional simulations that should take place before any firm conclusions on the river's ultimate quality include evaluation of: (a) current non-point

Table 4.8
Computer Simulation of Rio Grande Water Quality
(for flows of June 1988)

Parameter	No treatment	Primary treatment	Joint project
Dissolved oxygen minimum (mg/l)	6.3	7.0	7.8
BOD_5 maximum (mg/l)	5.0	3.0	1.2
Chloride minimum (mg/l)	206	206	206
Fecal coliform maximum (colonies/100 ml)	130,000	68,000	10

Source: Yang Ziming, LBJ School of Public Affairs, University of Texas at Austin, analysis of water quality data using Qual-TX simulation model, May 1, 1990.

discharges; (b) current low-flow and high wastewater discharge conditions; and (c) expected future low-flow and high discharge for the treatment plant's designed conditions (600,000 population).

In summary, conventional secondary wastewater treatment of Mexican municipal sewage appears to be sufficient to reduce contaminant discharges to levels that will not degrade the Rio Grande below Mexican or Texas surface water quality standards. Even if this is the case, there will remain a number of unresolved water quality conflicts between the United States and Mexico, as discussed in the section below.

Unresolved Water Quality Problems in the Rio Grande

The previous sections have concluded that (a) the Rio Grande has endemic quality problems deriving from inadequately treated wastewater and (b) simulations of conventional secondary wastewater treatment show resulting surface water quality meeting Mexican and Texas standards. There remain a series of issues that the United States and Mexico must address before the Rio Grande is again a river where fishing and swimming can be contemplated without health risks. These issues are:

1. The timing of wastewater treatment construction and the financing of plant operations and renovations.
2. Pollution sources which wastewater treatment plants would not reach.
3. The inherent awkwardness of binational programs.

It is not clear when or whether the required wastewater treatment plants will be constructed along the Rio Grande. Table 4.9 list the proposals in the current (1992) version of the Mexican-U.S. Integrated Border Environmental Plan.[48]

The plan does indicate that thought has been given to the problems in six of the problem areas along the Rio Grande/Río Bravo. However, most of the movement reflects plans rather than contracts to be implemented. The major success (the Laredo/Nuevo Laredo Project) was a product of local initiative supported by elected representatives. (See Chapter 5.)

Once decisions are made to build wastewater treatment plants, there remain questions about how infrastructure will be

Table 4.9
Water Quality Improvements in the 1992 Border
Environmental Plan in the Rio Grande Region

Twin city area	Planned infrastructure	Status
Ciudad Juárez/El Paso	Mexico will fund treatment system	Planned
Piedras Negras/ Eagle Pass	IBWC will develop a plan	Planned
Nuevo Laredo/Laredo	Secondary treatment plant under construction	Complete by 1994
	IBWC talks on industrial waste pre-treatment	Talks begun
Lower Rio Grande	Eliminate untreated sewage discharges	IBWC plan by 1992
Reynosa/McAllen	Increase collector system in Reynosa from 60 to 85%	Unscheduled
	Mexican side to recondition oxidation ponds	Complete by 1994
Matamoros	Mexican side to increase collector system from 65 to 75%	Unscheduled
	Mexican side to construct wastewater treatment plant	Unscheduled

Source: Jan Gilbreath Rich, "Planning the Border's Future: The Mexican-U.S. Integrated Border Environmental Plan," pp. 38-39.

maintained, operations will be funded, and renovation or improvements initiated. The joint Laredo/Nuevo Laredo Project did not indicate sources of funds to operate and maintain the plant, although cooperative financing is expressed as a goal.[49] That proposal did not address the issue of industrial pre-treatment of wastes that could damage the secondary treatment process, although this has become a matter of the bilateral agenda.[50] The timetable for cleaning up the Rio Grande will be long if each nation and the commission must address each step to expand or upgrade existing facilities only after the previous problem is resolved.

The issue of pre-treatment raises questions of coverage. Will existing plans take into account hazardous or toxic waste discharges? What will be done regarding non-point sources of pollution? What efforts can be made to abate the effluents from squatter settlements or industry outside the largest twin city areas that are not connected to municipal sewer systems?

Secondary wastewater treatment is not effective for handling wastes which can kill the bacteria which convert sewage into carbon dioxide and water. The large number of maquiladoras along the Rio Grande probably produce substantial volumes of water-borne residuals that are hazardous or toxic to people and other life. However, nine years after the La Paz Agreement, the two nations are still trying to determine how much waste is generated and where it goes.[51]

It is evident from physical evidence that a substantial volume of untreated organic waste is discharged into the Rio Grande from diffuse runoff. But only recently has the U.S. federal government considered studying the issue of non-point sources of organic pollution that degrade the quality of the Rio Grande.[52] That proposal is not yet funded.[53]

In a river basin within one political jurisdiction, the process of water quality management would follow a straightforward sequence of steps. The jurisdiction would establish standards, such as the quality of surface water for particular water uses and the volume/concentration limits on wastewater discharges. The jurisdiction would monitor surface water quality, the quantity/quality of point wastewater discharges, and measure/estimate non-point source pollutants. The jurisdiction would then plan, build, and operate any public infrastructure required to collect and treat point or non-point waste discharges.

As an international river, the Rio Grande/Río Bravo is subject to multiple jurisdictions, which has led to a fragmentation of responsibilities. At least two sets of ambient and effluent water standards apply, those of Texas and Mexico. No fewer than five agencies monitor water quality at various sites with diverse equipment. Construction is at least a three-part process, with each nation responsible for its own infrastructure but under binational supervision. Planning consists of collecting and arraying plans of federal, state, regional, and local governments along both sides of the Rio Grande.

This system of coordinated bilateral activity has led to the current situation. The Rio Grande is polluted and no one political jurisdiction is responsible for improving the environment.

This problem of management of the water issues along the Rio Grande/Río Bravo is the subject of the following chapter of this monograph. Water quality can be improved. With sufficient planning, money, and infrastructure, the river can become fishable and swimmable again. The issue is whether the two nations can find means for cooperating through joint activity rather than communicating about unilateral plans.

Notes

1. Penelope ReVelle and Charles ReVelle, *The Environment: Issues and Choice for Society,* third edition, Jones and Bartlett Publishers, Boston, 1988, p. 275.

2. David J. Eaton and John Michael Andersen, *The State of the Rio Grande/Rio Bravo,* The University of Arizona Press, Tucson, Arizona, 1987,. pp. 35, 40.

3. An analysis by Andrew Sherrill of water quality data for the period August 1, 1986 through July 31, 1989. Information derived from unpublished computer print-outs furnished by the Texas Water Commission's Statewide Monitoring Network, Austin, Texas, December 1989.

4. Texas Department of Water Resources, "Texas Surface Water Quality Standards," Report LP-71, Austin, Texas, April, 1981.

5. Telephone conversation between Donna Zinke and Osborne Linguist, International Boundary and Water Commission, March 27, 1989.

6. Secretaría de Agricultura y de Recursos Hidráulicos, *Reglamento para la Prevención y Control de la Contaminación de Aguas,* México, D.F., March 29, 1973.

7. SARH, *Reglamento.*

8. Andrew Sherrill, analysis of 1986-1989 water quality data.

9. Sherrill, analysis of 1986-1989 water quality data.

10. James E. Garcia, "Border River Laden with Wastes," *Austin American-Statesman,* Sunday, September 29, 1991, p. A17, Col. 3.

11. David L. Buzan, "Intensive Survey of the Rio Grande: Segment 2304," draft report, Texas Water Commission, Austin, Texas, June 1989, Table II.

12. David J. Eaton and John Michael Andersen, *The State of the Rio Grande/Río Bravo,* p. 169.

13. David J. Eaton and John Michael Andersen, The State of the Rio Grande/Río Bravo, p. 169.

14. Alfonso Ortiz, *Comparative Study of Infant Mortality in the Texas-Mexico Border Area of Laredo/Nuevo Laredo,* Working Paper No. 29,.Lyndon B. Johnson School of Public Affairs, Austin, Texas, 1984.

15. David C. Warner, "Health Issues at the U.S.-Mexican Border," *Journal of the American Medical Association,* Vol. 265, No. 9, January 9, 1991, p. 243.

16. Texas Department of Health, Bureau of Communicable Disease Services, "Texas Morbidity This Week," Austin, Texas, 1981.

17. Testimony submitted to the U.S. Environmental Protection Agency by Texas Attorney General Dan Morales on September 28, 1991 at El Paso, Texas. As quoted in Jan Gilbreath Rich, *Planning for the Border's Future: The Mexican-U.S. Integrated Border Environmental Plan,* Occasional Paper No. 1, U.S. Mexican Policy Studies Program, LBJ School of Public Affairs, Austin, Texas, February 1992, p. 21.

18. An analysis by Andrew Sherrill of wastewater discharge data for the first nine months of 1989. Data derived from unpublished Texas Water Commission self-reporting records of wastewater discharges for the Rio Grande Basin. Austin, Texas, December 1, 1989.

19. International Boundary and Water Commission, "Exhibit IV of the Joint Report of the Principal Engineers Concerning Measures that Should be Undertaken to Improve the Quality of the Waters of the Rio Grande at Laredo, Texas/Nuevo Laredo, Tamaulipas, Mexico," August 1989, p. 5; interview of Stephen W. Manning by Andrew Sherrill, April 6, 1990.

20. Unpublished letter from Narendra N. Gunaji (Commissioner, U.S. Section, International Boundary and Water Commission, El Paso, Texas) to Allen Beinke (Executive Director, Texas Water Commission, Austin, Texas), January 9, 1990.

21. Garcia, "Border River Laden with Wastes," p. A17.

22. The logic of discussion of the international law and water quality in the Rio Grande Basin was developed in Donna Patricia Zinke, "The Legal Regime Governing Pollution of the Rio Grande," Professional Report, Lyndon B. Johnson School of Public Affairs, Austin, Texas, May, 1989.

23. Nisuke Ando, "The Law of Pollution Prevention in International Rivers and Lakes," in the *Legal Regime of International Rivers and Lakes,* Zacklin and Caflisch, editors, Boston: Nijhoff, 1981, p. 333.

24. Albert E. Utton, "International Water Quality Law," in Teclaff and Utton, editors, *International Environmental Law,* 1975, p. 154.

25. Albert Utton, "International Water Quality Law," pp. 159-160.

26. Anne M. Morgan, "Transboundary Liability Goes with the Flow, Gasser v. United States, The Use and Misuse of a Treaty," in *Transboundary Resources Report,* Vol. 5, No. 3, (Summer 1991), p. 3.

27. International Law Association, Helsinki Rules, 1966, Article IV; Ando, "Pollution Prevention," p. 345; Report of the United Nations Conference on the Human Environment, United Nations, New York, 1973; Stephen C. McCaffrey, "The Work of the International Law Commission Relating to the Environment," *Ecology Law Quarterly,* Vol. 11, 1983, p. 189; and Jens Evensen, "Second Report on the Law of the Non-Navigational Uses of International Watercourses," April 24, 1984.

28. Lawrence A. Herzog, "Transfrontier Ecological Planning: Some Lessons from Western European-United States-Mexico Border Region Comparisons," *Transboundary Resources Report,* Vol. 3, No. 2, (Summer 1991), p. 7.

29. "Utilization of Water of the Colorado and Tijuana Rivers and of the Rio Grande Treaty Between the United States of America and Mexico," November 14, 1944, "Treaty Series 994, Article 3, U.S. Government Printing Office, Washington, D.C., 1946, p. 8.

30. Stephen P. Mumme, "The Background and Significance of Minute 261 of the International Boundary and Water Commission," *California International Law Journal,* 1981, p. 223.

31. Stephen P. Mumme, "International Boundary Water Commission," *Transboundary Resources Report,* Vol. 1, No. 1, Spring 1987, p. 6; and "Minute Order No. 279," International Boundary and Water Commission, Laredo, Texas and Nuevo Laredo, Tamaulipas, August 28, 1989.

32. Rio Grande Pollution Correction Act of 1987 (H.R. 2046), 100th Congress, First Session, April 9, 1987 (signed October 3, 1988).

33. "Agreement Between the United Mexican States and the United States of America on Cooperation for Protection and Improvement of the Environment in the Border Area," La Paz, Baja California Sur, Mexico, August 14, 1983.

34. U.S. Environmental Protection Agency and the Mexican Secretaría de Desarrollo Urbano y Ecología, *Integrated Border Environmental Plan for the Mexican-U.S. Border Area, (First Stage 1992-1994),* working draft, August 1991.

35. James E. Garcia, "Border River Laden with Wastes."

36. U.S. International Trade Commission, *Production Sharing: U.S. Imports under Harmonized Tariff Schedule Subheadings 9802.00.60 and 9892.00.88, 1987-1990*, Washington, D.C., 1991, pp. 8-20.

37. "Utilization of Water," p. 8.

38. "Utilization of Water," pp. 13-14.

39. International Boundary and Water Commission, "Minute Order No. 279," Resolution 6, Laredo, Texas and Nuevo Laredo, Tamaulipas, August 28, 1989, p. 4.

40. U.S. EPA and the Mexican SEDUE, *Integrated Border Environmental Plan* , working draft.

41. 31 Texas Administrative Code Section 307. Additional information from Richard Kiesling, Texas Water Commission, telephone interview with David Hurlbut, July 28, 1992.

42. International Boundary and Water Commission, "Joint Report of the Principal Engineers Concerning Measures That Should Be Undertaken to Improve the Quality of the Rio Grande at Laredo, Texas/Nuevo Laredo, Tamaulipas," Ciudad Juárez, August 25, 1989, p. 5.

43. IBWC, Exhibit IV, August 25, 1989.

44. IBWC, Exhibit IV, August 25, 1989.

45. David L. Buzan, "Intensive Survey of the Rio Grande Segment 2304," Draft Report, Texas Water Commission, June 1989.

46. IBWC, Joint Report of the Principal Engineers.

47. Ziming Yang, unpublished report of results of simulations of Rio Grande water quality, May 1, 1990.

48. Jan Gilbreath Rich, "Planning for the Border's Future," pp. 38-39.

49. IBWC, "Joint Report of the Principal Engineers," p. 11.

50. Jan Gilbreath Rich, "Planning for the Border's Future," p. 39.

51. Jan Gilbreath Rich, "Planning for the Border's Future," pp. 34-35.

52. U.S. Bureau of Reclamation, "Plan of Study: Lower Rio Grande Basin Study—Texas," U.S. Bureau of the Interior, December 1991, unpublished proposal, p. 11.

53. Unpublished letter from Fred R. Ore, U.S. Bureau of Reclamation, to Max Sherman, Dean, Lyndon B. Johnson School of Public Affairs, December 31, 1991.

Chapter 5:
Institutional Solutions

THE PREVIOUS CHAPTERS HAVE DOCUMENTED THE SURFACE water allocation, water quality control, and groundwater management problems that exist within the Rio Grande/ Río Bravo Basin. Only one institution has been created to resolve these or other water issues: the International Boundary and Water Commission (IBWC)/La Comisión Internacional de Límites y Aguas (CILA). The IBWC/CILA is the shared international body constituted in two cooperating national sections established by treaty to manage water and land boundary issues within the limitrophe parts of Rio Grande/Río Bravo.[1] Mexico and the United States have given the IBWC/CILA jurisdiction that is exclusive of and superior to any federal, state, or local institution. Thus, the hopes and frustrations of border residents for solving water conflicts along the Rio Grande/Río Bravo have been focused on the IBWC/CILA. In 1983, when Mexico and the United States created a new mechanism for cooperating on border environmental problems through the Environmental Protection Agency (EPA) and the Secretaría de Desarrollo Urbano y Ecología (SEDUE), the agreement stipulated that "nothing in this Agreement shall prejudice or otherwise affect the functions entrusted to the International Boundary and Water Commission."[2] As the two nations developed an integrated environmental plan for the Mexico-U.S. border area, the IBWC/CILA remains the institutional and budgetary vehicle for processing water allocation and water quality management infrastructure change.[3]

The role of EPA and SEDUE in carrying out the border environmental plan has sparked a public debate on the role of the IBWC/CILA in implementing border environmental programs.[4] This chapter focuses on one part of the public debate—whether institutional alternatives exist to the IBWC/CILA for water management along the Rio Grande/Río Bravo.

Five separate mechanisms for water resource decision making can be identified by examining historical and current water management practices for the Rio Grande/Río Bravo. (See Table 5.1.) All five are legal, all accept the legitimacy of national sovereignty and prior binational treaties, and four of the five are rooted in recent experience. If the discussion shifts from the use of the IBWC/CILA versus the EPA/SEDUE as lead agencies for riparian control, a variety of institutions can be identified as working concurrently to improve the surface water allocation, water quality control, and groundwater management along the Rio Grande/Río Bravo.

Origins of Bilateral Cooperation

The Peace Treaty of 1848 settling the Mexican and American War established the Rio Grande/Río Bravo as the boundary between Mexico and the United States.[5] As early as 1853 the U.S. and Mexico jointly created temporary commissions to survey and demarcate the boundary on the ground to deal with problems that arose whenever the course of the river moved from time to time.[6]

On March 1, 1889, the governments of Mexico and the United States created the International Boundary Commission (IBC), consisting of Mexican and U.S. sections. The IBC was charged with enforcing treaty rules as well as settling questions regarding the boundaries when the rivers changed course.

As part of the 1944 water treaty, Mexico and the United States agreed to resolve border water and sanitation problems through a modified joint body, the International Boundary and Water Commission (IBWC) composed of Mexican and U.S. Sections.[7] The treaty assigned to IBWC the task of assessing the feasibility of building hydroelectric facilities, storage dams, or flood control projects. The treaty stipulated that matters concerning joint action be handled through the U.S. Department of State and the Mexican Secretariat of Foreign Relations.[8] The treaty also creates an IBWC procedure to develop new infrastructure projects or policies not included in the Treaty of 1944. The decisions or recommendations of the two governments are recorded as minutes, which are signed by each of the two commissioners.

Once approved by each government, minutes become binding upon the two governments. For example, the U.S. Congress

Table 5.1
Alternate Models for Rio Grande Water Management

Initiator of action	Role of IBWC/CILA	Illustration
IBWC/CILA	national policy initiator	dam construction along the Rio Grande
federal government	federal policy implementor	Integrated Environmental Plan for the Mexico-U.,S. Border Area
state or local agenwater agencies	communications facilitator	water rights trading in Rio Grande/Texas
local governments	regional coordinator	Laredo sanitation and treatment project
regional binational private water utility	quality controller	private regional hydraulic utility

Source: U.S.-Mexican Policy Studies Program, LBJ School of Public Affairs.

would implement a cooperative project through normal channels of authorization and appropriation. The U.S. and Mexican sections work through the respective agencies or institutions within their own governments, so the principle of national sovereignty is not compromised. Each nation operates and maintains the part of the projects assigned to its section. To undertake these tasks, the Treaty of 1944 stipulates that an Engineer Commissioner with diplomatic immunity head each section and that each commission may employ staff "as it deems necessary."[9] In 1990, the U.S. Section employed about 260 people and the Mexican Section had 92.[10]

Consultants or engineers attached to each section prepare studies and reports prior to any joint project or agreement. After the initial work is completed, individual section engineers report recommendations back to their own respective section. Only after both sections have been briefed by engineers does the proposal go before the Mexican and U.S. engineer commissioners. When the commission meets to hear or discuss proposals, the agreements are transcribed in the form of minutes. Once both sections' commissioners have signed, minutes are as binding as the provisions of the water treaty.

Although the Treaty of 1944 designates IBWC/CILA as the only international organization to deal with conflict resolution and cooperative development in the Rio Grande/Río Bravo Basin, other federal and state agencies within each nation have overlapping duties. Former Mexican Section Commissioner Joaquin Bustamante has stated that jurisdictional overlaps exist among the duties of CILA, the Secretariat of Foreign Relations (under whose responsibilities the Mexican Section falls), SEDUE, and the Secretariat of Agriculture and Hydraulic Resources.[11] The U.S. Department of the Interior's Geological Survey and Bureau of Reclamation sometimes are consulted prior to decisions and projects. Within Texas, any U.S. Section agreement or proposal that might affect Texan water rights or laws would be cleared with the Texas Water Commission (TWC). However, IBWC directives take precedence. For example, the TWC Watermaster requests releases of water from the IBWC, and the IBWC controls the rate of discharge and flow. When IBWC samples river water quality, the samples are sent to the TWC to be analyzed, and results are shared by the IBWC with the Mexican Section.[12]

The IBWC as Policy Initiator

The IBWC/CILA quickly gained a reputation as a commission that could develop national priorities within its limited jurisdiction and build infrastructure.

The Treaty of 1944 directed the IBWC/CILA to jointly construct dams required for the conservation, storage, and regulation of water within the basin, and determine the most feasible sites, the maximum reservoir capacity, the conservation capacity required by each country, the capacity required for silt retention, and the capacity required for flood control.[13] The Treaty also stated that "the Construction of the international storage dams shall start within two years following the approval of the respective plans by the two Governments. The works shall begin with the construction of the lowest major international storage dam."[14]

Joseph F. Friedkin, former U.S. Section Commissioner, said that the Mexican and U.S. engineers quickly agreed on the site for the location of the lowest dam, "Falcon was as far downstream as politically possible, and would be effective in terms of flood control."[15] The United States agreed to pay for a 58.6 percent share of the Falcon Dam's storage capacity, while the Mexican side paid for 41.4 percent. The U.S. Congress authorized the construction, operation, and maintenance of Falcon by the American-Mexican Treaty Act of September 13, 1950. The U.S. Department of the Interior and the Bureau of Reclamation planned the design of the dam. The U.S. section initiated construction later that year and the IBWC monitored and supervised each phase of the project. Falcon Dam was jointly dedicated on October 19, 1953 by the U.S. and Mexican Presidents.[16] The two countries also agreed to build an electricity generation plant to produce power for the region. Project costs and benefits were determined by the 1944 treaty: "Each government shall pay half the cost of the construction, operation and maintenance of such plants, and the energy generated shall be assigned to each country in like proportion."[17]

In its first year of operation in 1954, Falcon Reservoir completely contained the highest flood of record in the Rio Grande; this fortuitous event confirmed the value cooperative development between the United States and Mexico.[18] The severity of the 1954 flood demonstrated to both sections that additional storage was needed to regulate the flow of the waters and that Falcon was too far downstream to provide sufficient capacity to store waters for the entire basin.[19] IBWC studies established that a second dam upstream would enhance Falcon, as both could operate for flood control, coordinating releases and storage.

IBWC engineers investigated approximately 30 sites (particularly around the Big Bend area) prior to selecting a location just below the confluence of the Pecos and Devils Rivers.[20]

The Amistad (Spanish for friendship and a translation of the Indian word Tejas) Dam and Reservoir project was designed jointly by the engineers of the respective sections and cooperating agencies in conjunction with separate contractors. The U.S. part of the dam was designed by the U.S. Army Corps of Engineers, Southwest Division. In accordance with the treaty, each government independently determined exactly how much storage capacity it would need from the dam. As the United States and Mexico agreed to split the capacity into 56.2 and 43.8 percent shares, respectively, the total cost of the construction of the dam was divided in this manner as well. U.S. President Dwight Eisenhower and Mexican President Adolfo López Mateos met and agreed on dam construction in October, 1960.[21] Each side of the Amistad Dam was built separately. The construction crew of each side tried to outperform each other to build its part faster and more efficiently. U.S. President Richard M. Nixon and Mexican President Díaz Ordaz dedicated the Amistad Dam in September 1969. Both sections of the IBWC jointly operate and maintain the dam. Water is released according to each country's agreed allocation as well as downstream requirements. For the United States, downstream requirements are determined by the Texas Water Commission.

The Falcon and Amistad Dams and Reservoirs established the commission as a policy initiator. Given a limited mandate by the Treaty of 1944, the two sections demonstrated that they could jointly (a) identify and specify needed water infrastructure projects, (b) convince each national legislature and executive to finance its portion, (c) implement and oversee construction in a timely and cost-effective manner, and (d) maintain and operate the facilities to mutual benefit over time. As a result, the IBWC/CILA became a model agency for other countries, showing how two nations could act on cooperative projects to achieve benefits that neither side could attain alone.

The Commissioner concept developed in 1928 when Mexico and the United States created a committee of three engineers from each side to study water distribution within the Rio Grande and the Colorado Rivers. This committee concluded that a bilateral organization should be created to allocate available resources. This group of engineers, joined by others, including a few legal experts, were responsible for drawing up the 1944 treaty. Cruz Ito, Engineer Assistant to U.S. Section Commissioner, believes that the rapid and sustained success derives from the structure of the IBWC/CILA as a commission composed of engineers concerned with the specific issues identified by the water treaty or subsequent minutes.[22]

The IBWC as Implementor of Federal Agency Priorities

In 1983, Presidents Reagan and de la Madrid established a new route for bilateral communication through the 1983 Agreement on Cooperation for the Protection and Improvement of the Environment in the Border Area. While the 1983 agreement, named for the city of La Paz, Baja California Sur where it was signed, confirms the unique jurisdiction of the IBWC/CILA; it also authorizes the heads of the respective federal environmental agencies—EPA and the new Secretaría de Desarrollo Social (Sedesol, which replaces SEDUE)—to communicate with each other and develop cooperative plans and programs to maintain and improve environmental quality along the border. In 1992, EPA/SEDUE produced the final version of the "Mexican-U.S. Integrated Border Environmental Plan."[23] Within the context of the La Paz agreement and the 1992 plan, the IBWC/CILA has been given a charge to implement federal infrastructure policy along the Rio Grande/Río Bravo.

The binational border environmental plan[24] includes a list of planned and possible agency expenditures that reflect the EPA/Sedesol priorities for unilateral but coordinated binational efforts to improve the border environment. Water infrastructure represents the top priority in terms of actual committed funds, and the Plan states that the IBWC/CILA should continue to be the primary vehicle for water infrastructure construction.[25] Although the IBWC had a clear role in identifying projects and expenditures, it is not the lead institution for determining binational environmental quality issues. On the U.S. side, the EPA establishes priorities and on the Mexican side, Sedesol does the same. On the U.S. side, the Executive Branch allocates tasks among agencies and Congress authorizes funds to implement those priorities. The IBWC/CILA has become the agency of choice to supervise water infrastructural programs which cannot be undertaken by one side or the other and where coordinated construction and operation are central to project success.

The IBWC/CILA's future role as implementor of federal plans is unclear. For example, the U.S. section, as a quasi-independent unit within the U.S. Department of State, is at a disadvantage vis-à-vis the Executive Branch agencies because the IBWC is not able to lobby Congress for projects or funds. It is an agency designed to be outside the interagency councils that set priorities and policies.

For the EPA/SEDUE integrated environmental plan, the IBWC/
CILA may remain as constructors of binational infrastructure.
While the Commission will be neither silent nor devoid of preferences, it is now and likely shall continue to be limited in in-
dependence of action or capacity to influence broader federal
border infrastructure priorities.

State Initiatives: The IBWC as Facilitator

As an international agency the IBWC/CILA works through
national and sub-national institutions on each side of the border. State, regional, or local institutions can originate water
management or infrastructure projects which remain on one
side of the border, and coordinate with parallel activity on the
other side. Even when the initiative, funds, plans, and management come from other institutions, the IBWC/CILA can still ful-
fill its treaty role to study, investigate, and communicate about
Rio Grande/Río Bravo water resource issues.[26] For example, the
state of Texas now encourages water rights transfers that benefit both nations and supports border water infrastructure construction, while the IBWC has no role except to communicate
with the Mexican side.

In 1991, Texans voted to approve a bond package that directed $150 million into a revolving loan fund for use in infra-
structure improvements in Texas colonias.[27] An estimated
300,000 Texans live in these squatter settlements, where hous-
ing may lack internal plumbing and public water supply or
wastewater drainage may be missing.[28] Although such settlements line both sides of the Rio Grande/Río Bravo, partly a response to the increase in industrial activity along the border,
there has been no binational effort to upgrade colonia infrastructure. So far, 19 water and wastewater service projects to-
taling $110 million are under construction. Supplementing these
state efforts is a $15 million federal loan program to help colonia
residents install plumbing that will connect them to the community systems.[29] In addition, the border environment plan calls
for $5 million in federal funds, $50 million from the EPA and
$25 million from the U.S. Department of Agriculture.[30]

The Texas state funds allocated for colonia water and wastewater infrastructure (augmented by U.S. federal agency funds)
are not sufficient to resolve the colonias' problems, but are a

first step in what could become a binational effort to upgrade living conditions for the poor along the border to levels consistent with the remainder of the state. The leadership is coming from Texas government and local water/wastewater utilities—not from federal agencies or the IBWC/CILA. This state initiative model is an alternative to either an international effort (through the IBWC/CILA) or coordinated national activities (through EPA and Sedesol).

Texas is also attempting to improve water supply for border twin cities with water rights trading coordinated through the Texas Water Commission. As a result of a 1956 court case (the State of Texas vs. Hidalgo County Water Control and Improvement District No. 18), the Texas Legislature passed the Water Rights Adjudication Act of 1967. This act led to the adjudication of all Texas water rights along the Lower Rio Grande in 1971.[31] The Rio Grande Watermaster's Office in the Texas Water Commission initiates water withdrawals for Lower Rio Grande users through the IBWC and can divide "water of streams or other sources of supply in his division in accordance with the adjudicated water rights."[32] In times of drought, the watermaster may reallocate water to users based on existing water rights.[33]

Chapters 2 and 3 indicated that conflicts may develop between Texas and Mexico over surface water allocation and groundwater withdrawals. One key pressure point in the system is the increase in demand for municipal water in both Mexican/Texas twin cities reflecting increased population and economic activity. The Lower Rio Grande Watermaster could broker water rights trades along the U.S. side of the border to ease those pressures, by reallocating under-utilized agricultural water rights for municipal uses. A pattern of unilateral reallocation of water rights along the border—separately within the U.S. and within Mexico—can provide water flexibility for accommodating economic changes without requiring treaty renegotiation.

The cities of Harlingen, Brownsville, and McAllen have been able to purchase additional water and water rights from agricultural users through the cooperation of the Texas Water Commission. The city of McAllen, Texas purchased water rights to 11,250 acre-feet from the United Irrigation District of Mission, Texas in 1989. The cities of Harlingen and Brownsville have also converted agricultural water rights into municipal water rights through purchase. In addition, Brownsville and McAllen have obtained raw water from irrigation districts and municipalities to augment their supplies.[34] When Texas municipalities can obtain new secure water through purchase, pressures are reduced for either trade in water across the border or for enhanced groundwater extraction. While the size of water-rights trades are not yet large, the principle has been established that water rights can be converted from agricultural to municipal and commercial uses.

Two keys to this successful shifting of water rights to a higher valued use are the certainty regarding water rights ownership and the active involvement of the Texas Water Commission in providing legal transfer of water rights through the allocation programs of the watermaster. Prior to 1992, only the Lower Rio Grande was actively managed by a watermaster.[35] Water rights marketing among users within each nation can provide both greater flexibility and easier administration in responding to changing water requirements than either new infrastructure construction or renegotiation of water rights between Mexico and Texas. By their nature, such programs will not involve international or federal institutions on the U.S. side, except for the federal roles in communicating the changing water use patterns to Mexican colleagues.

Local Initiative

The Laredo/Nuevo Laredo Water Sanitation and Treatment Plant represents local as well as a binational initiatives for improved wastewater infrastructure, as mediated through state and federal government agencies. This concept of a local initiative–resulting in eventual IBWC/CILA management of a project jointly funded by federal, state, and local sources—offers a new administrative alternative for improvement of water or wastewater infrastructure along the Rio Grande/Río Bravo.

Minute Order 279 of August 28, 1989 proposed that the United States and Mexican governments jointly construct an international sanitation project in Nuevo Laredo to improve the quality of the waters of the Rio Grande. The project is designed to "stop the discharge of untreated sewage into the Rio Grande from Nuevo Laredo" and provide sufficiently treated sewage to meet U.S. effluent standards.[36] CILA planned the construction of the plant, and the IBWC is supervising the construction, operation, and maintenance of the facilities. Texas, the United

States, and Mexico will each participate in financing construction of this project.

Minute Order 279 addresses the continuing problem of untreated sewage entering the Rio Grande from Nuevo Laredo, Tamaulipas. Nuevo Laredo, with an estimated 1988 population of 400,000, discharges an estimated 876 liters per second (20 million gallons per day) of untreated sewage from some 28 locations.[37] Nuevo Laredo does not treat its wastewater along this segment of the Rio Grande, but discharges sewage directly from downstream drains, including three drains upstream from Nuevo Laredo's water supply. Without the new plant, the volume of sewage would increase to about 31 million gallons per day by the year 2000 and future discharge points for Nuevo Laredo would be upstream from the water supply for the City of Laredo, Texas.

In 1979, Mexico constructed chlorination facilities at two sites in response to the U.S. Section of the IBWC identifying the two major drains in Nuevo Laredo discharging untreated sewage to the river. However, the plants never opened or operated because Mexico did not have the funds for a sewage collection system and treatment plant.

Article 3 of the 1944 Water Treaty (59 Stat. 1212), authorizes U.S./Mexican actions "to give preferential attention to the solution of all border sanitation problems. . . . "[38] Such action is consistent with the provisions of the La Paz agreement.

Joint projects include binational construction of an improved collection system in the city of Nuevo Laredo; new sewage collectors; and a 31-million-gallon-per-day secondary treatment plant in Mexico, approximately 11 kilometers (7 miles) downstream of the Juárez/Lincoln International Bridge between Laredo and Nuevo Laredo. The United States and Mexico each agreed to pay half of the cost of $22 million. The state of Texas will contribute an estimated $2 million to the U.S. share. The two countries will share the operation and maintenance costs.

Both nations intend for the discharge from the proposed Nuevo Laredo treatment plant to meet U.S. effluent standards and the EPA's minimum surface water quality standards. By signing the agreement, Mexico agreed to prevent discharge of untreated wastewater into the Rio Grande, as well as to "assure efficient operation of the treatment plant."[39] A do-nothing approach would result in continued reduction of water quality and enhance the risks of waterborne communicable diseases on both sides of the River. In addition, water quality in the Falcon Reservoir would be diminished due to further pollution.

The rationale for a bilateral sewage treatment plant based in Nuevo Laredo is that any unilateral decisions by either side would not resolve the Rio Grande/Río Bravo water quality problem below the twin cities. An international wastewater treatment plant designed to handle the City of Nuevo Laredo's sewage load built in the United States would be more expensive than one in Mexico, and would be less likely to collect Mexican sewage.

The intent of the proposed joint plant is to rehabilitate Nuevo Laredo's sewage collection system, provide secondary sewage treatment before discharge, and eventually stop increasing discharges of untreated sewage into the Rio Grande. The proposed 31 million-gallon-per-day treatment plant has a capacity 50 percent larger than the level of existing discharges.

A number of specific beneficial effects on the environment will result in the Laredo, Texas area. The reduction in discharge of untreated wastewater should improve health in Mexico and Texas. A report of the U.S. General Accounting Office on health problems points out that diseases such as amoebiasis and shigellosis, which are caused by unsanitary conditions, are prevalent along the border.[40]

The improvements in Rio Grande/Río Bravo sanitation should improve the economics of the border area of Laredo, Texas and Nuevo Laredo, Tamaulipas. Cleaner water in the Rio Grande/Río Bravo would improve the business climate through the reduction in water-borne diseases and the renewed quality of the environment. The IBWC proposal may allow Mexico to increase industrialization in the area without increasing raw sewage discharges in the Rio Grande. Currently, about 70 maquiladoras (mostly U.S.-owned assembly plants) are located in the Nuevo Laredo area, a significant increase over the nine plants located there in 1983.

The Laredo/Nuevo Laredo Sanitation Project began as a local initiative and was eventually processed through the U.S. Congress to the IBWC/CILA. The impetus for it began with the Laredo Water Works System's legal requirement to maintain surface water quality standards set by the U.S. Environmental Protection Agency. The Water Works System is a public agency charged with providing water services to the Laredo area. Fernando Roman, superintendent of wastewater, points out that "for the longest time we were beyond the [EPA] standards on account of a very reduced treatment capacity."[41]

In 1973, Del Mar, a city north of Laredo, was ordered by a U.S. federal court to eliminate untreated sewage discharges into the Rio Grande. As director of the Del Mar Water District (now assistant director of Laredo Water Works System), Tomás Rodriguez hoped to change Del Mar's discharge status. "We invested $20,000 to pump the effluent someplace else. The city of Laredo didn't do anything at the time. Then they hired Howard Gaddis."[42]

With Gaddis as its director, the Laredo Water Works System studied its facilities and concluded it did not have the capacity to treat the wastewater generated by Laredo. The Laredo Water Works Board authorized Gaddis to design a new wastewater treatment plant. In the meantime, the City of Laredo was preparing to acquire the Laredo Water Works System, which was a separate public entity.

With the backing of the board, Gaddis developed the construction plans for a new plant. However, as Fernando Roman explains, "the EPA was granting at the time 55 percent of the construction money. The consulting engineer and Gaddis could not meet the deadline to submit the application so we lost the opportunity to get $70 million of free money."[43] Gaddis was subsequently fired because the Water Works had paid $800,000 for the design and had no chance of building a plant.

In 1982, under the leadership of J. R. Mathis, the Laredo Water Works System identified the quality of the Rio Grande as the primary local source of concern over pollution and health. At a 1983 meeting of the two Laredos, an international committee comprised of the heads of various city departments from the cities of Laredo and Nuevo Laredo, Mathis and Roman were able to quantify Nuevo Laredo's discharges. Roman said that Nuevo Laredo was dumping 22 million to 23 million gallons of sewage each day into the Rio Grande. He commented that:

> Everyone was saying 'we can't do anything about it; there's no money, there's not an infrastructure.' And the American side couldn't say anything either. Then J.R. Mathis said . . . 'I had a dream that one day these two countries will get together to solve their sanitation problems. I would like to propose that we build a joint sewage plant.'[44]

This rather spontaneous proposal proved to be the turning point of wastewater treatment along the Rio Grande. At the time, Laredo hoped to construct a plant on the U.S. side to treat both cities' discharge at the same location. One selling point for the Mexicans was the fact that Laredo had already paid for planning and designing the system. In addition, each side would have to pay only half of the construction and maintenance costs.

Laredo residents were excited about the proposal. Roman said that "the idea was in the newspapers. We started to get a lot of callers wanting to know about this project."

There was some ambivalence about the project because of rising water rates. In the early 1970s, the Laredo Water Works sold bonds to renovate its water treatment plants. To repay the bonds, the Water Works raised water rates. In 1983, the base water utility rate (the charge for a consumer using the base amount of water a month), rose from a minimum of $1.25 to a minimum of about $3.80 a month to pay for the water treatment plant. The average ratepayer's bill for sewage services was around $17 to $18.[45]

A drought hit the Laredo/Nuevo Laredo area in the summer of 1984. Roman was deluged with phone calls from residents of Nuevo Laredo asking why there was a water shortage. The residents of Nuevo Laredo did not understand why Laredo had enough treated water to sustain them through the drought, while their city was short on water and long on pollutants. Nuevo Laredo was still operating an antiquated water treatment plant.

Roman believes that Nuevo Laredo was "treating one part of the polluted water and blending it with the other part which was untreated or partially treated." In many parts of Nuevo Laredo, this partially treated water was the only water available. Even if the water was not used for drinking, it was used to wash dishes or floors and for showers, prompting a hepatitis epidemic. About 200 to 500 hepatitis cases were reported on the U.S. side, "but across the river the number was in the thousands. If you went into the hospital with a broken leg, you came out with hepatitis."[46]

The 1984 drought taught the people of Nuevo Laredo that water quality in the Rio Grande/Río Bravo could affect their drinking water system. Many had visited Laredo or sent their children to U.S. schools. They noticed the higher water quality in Laredo. They noticed that when people from Laredo crossed the bridge into Mexico they drank only bottled water. Many residents on the Mexican side began complaining to their local government.

The Mexican government was aware of the problem but the local government could not act independently of federal control. Roman said that "men from Mexico City" came to the border to perform tests and studies on the river and water, without contacting or meeting the local officials in Nuevo Laredo. The expense of a new treatment plant caused the Mexican federal government some concern. The federal government would have to pay the Mexican share, because the state of Tamaulipas and city of Nuevo Laredo did not have the tax base necessary for the project. Indeed, Laredo officials reported that Nuevo Laredo does not charge enough to cover expenses of existing water services and does not aggressively collect fees due.

To build the project on Mexican soil and under Mexican jurisdiction, the IBWC agreed to manage the project, deferring to the Mexicans on control of the construction process. Under this approach, the U.S. can inspect the facility periodically to ensure that the wastewater treatment plant functions according to the agreement. Each side will pay equal shares of the operational and maintenance costs.[47] Even after informal agreement among the IBWC, CILA and the Laredo/Nuevo Laredo water utility staffs, there remained the awkward problem of fund raising.

On the U.S. side, U.S. Sen. Phil Gramm (R-Texas), U.S. Rep. Kika de la Garza (D-Texas), and U.S. Rep. Albert Bustamante (D-Texas) lobbied for the project in Congress. Representative de la Garza attempted to interest the EPA and the Department of the Interior in developing an administrative remedy. According to de la Garza, the EPA had the authority to assist with pollution control along the Rio Grande but was unwilling to take on the responsibility because of the jurisdictional issues relating to Mexico and the IBWC.[48]

De la Garza created what became the Rio Grande Pollution Correction Act.[49] The bill authorized the U.S. Secretary of State, acting through the IBWC Commissioner, to carry out agreements with the Mexican government to collect and treat inadequately treated wastes in the twin cities along the Rio Grande. This meant that the United States would jointly fund waste treatment projects with Mexico that would deal solely with Mexican sewage treatment.

Congressman de la Garza's bill (H.R. 2646) passed the House

of Representatives on October 19, 1987. The bill was signed into law on October 3, 1988.[50] The law authorizes the U.S. Secretary of State to recommend the construction, operation, and maintenance of sewage treatment plants after consultation with the administrator of the EPA. The law establishes a formula for division of costs between Mexico and the United States and funds the U.S. share through congressional appropriations.[51]

The Laredo/Nuevo Laredo Sanitation Project is but one example of the involvement of state, federal, and international agencies after local water utility officials have reached informal bilateral consensus. A recent paper regarding wastewater treatment in Nogales, Sonora and Nogales, Arizona describes a similar process of local initiative and IBWC/CILA response.[52] It is likely that local initiatives in combination with multi-jurisdiction funding (at the local, state, and federal levels) will become a major vehicle for initiating water and wastewater infrastructural improvements along the Rio Grande/Río Bravo.

Hydraulic Regionalism

One final alternative administrative route for initiating water infrastructural improvements along the Rio Grande/Río Bravo can be called hydraulic regionalism: creation of private binational water utilities controlled by local governments in both Mexico and Texas to manage surface water use, groundwater withdrawals, and wastewater treatment utility. While no binational private but publicly controlled water utilities now exist along the border, there is ample legal and administrative precedent for them. Operation of such utilities need not challenge the jurisdiction of the IBWC/CILA or create sovereignty problems.

Poor communities tend to lack the expertise, financial resources, and stability to undertake development, management, and improvement of sophisticated water supply, wastewater treatment, or groundwater management projects. Traditional institutional remedies, such as reliance upon private engineering consultants or central government provision of services have a mixed record of success in areas where there are large numbers of underserved citizens.[53] Regional or multi-community utilities (which can be private, non-governmental institutions controlled by local governments) can develop the

geographical scale, the internal expertise, the financial indepen-
dence, and the levels of community support necessary for successful, reliable, and permanent provision of water and
wastewater services to poor communities. A regional entity can
enhance the development of community infrastructure, train
and employ local citizens, and respond to local community
priorities.

The best example of regional water infrastructure utilities is
the case of provincial water companies in the rural Netherlands
during the period of 1920 to 1960.[54] Provincial water utilities
were developed in seven Dutch provinces through a combination of local initiative, affordable water rates, and partial government subsidies. Each regional water company is a
self-financing and private entity managed by a board selected
by the local and/or provincial governments. Each developed
from an impetus of local residents, assisted by a plan generated
by the central government. Each provides water services in urban and rural areas and is able to take advantage of economies-
of-scale for water extraction, treatment, and distribution. Each
has used part of its income to finance continual expansion of
the distribution network into unserved rural areas. The pricing,
personnel, and management policies have fostered a loyalty and
a sense of participation among users, employees, and local
elected officials. These regional companies have gone beyond
delivery of water to facilitate general infrastructure improvements in the regions they serve. They have promoted modern-
ization in water-related industries. Each company has been a
steady employer of unskilled and semi-skilled labor in regions
where unemployment levels have been relatively high. Each has
remained solvent and attractive to bond investors, as fees were
designed to assure efficiency, equity, and return on invested
capital.

Hydraulic Regionalism on the Texas/Mexico Border

Whether the issue is water supply, groundwater management,
or wastewater treatment, the primary focus of attention is on
the three sister metropolitan areas of Ciudad Juárez/El Paso,
Nuevo Laredo/Laredo, and Matamoros/Brownsville. Secondary
problems are likely to occur in the sister cities of Piedras
Negras/Eagle Pass, Ciudad Acuña/Del Rio, and Reynosa/

McAllen. Each of these regions faces an unique set of circumstances that affects both the Mexican and Texas sides of borders. Each area expects to expand its regional economy and diversify employment opportunities for citizens as its population expands during the coming decades in response to the integration of the border economy. Each sister city pair has a large fraction of the population with incomes below the U.S. poverty floor.

The hydraulic regional concept could be applied in each of these binational sister cities by establishing a private utility with a service region encompassing the entire metropolitan area on both sides of the border. The scope of each utility's services could include surface water use, groundwater withdrawals, drainage, and wastewater treatment. These water infrastructure problems would be defined as a regional responsibility, rather than a set of independent problems for each of the many existing private and public water supply, wastewater treatment, and drainage utilities.

Each of the six regional hydraulic utilities would establish a goal of technical self-sufficiency. The utility staff should be capable of doing all technical tasks necessary to plan for, construct, operate, maintain, repair, and improve water infrastructure. These tasks include at least the following roles:

1. Geophysical survey and hydrological testing.
2. Planning for groundwater and surface water extraction.
3. Design of water treatment and wastewater treatment plants.
4. Development of plans for water distribution systems as well as drainage and sewerage networks.
5. Construction and supervision of construction of the pipe networks and treatment plants.
6. Laboratory assessment of groundwater and surface water quality and achievement of wastewater and drinking water standards of both Mexico and Texas.
7. Routine maintenance and repair of pipes, pumps, electrical equipment, computerized control equipment, and meters.

The appropriate skills would be developed primarily through apprenticeship, so as to recruit as many staff members as possible from unskilled or semi-skilled residents of the area. Staff would be trained through on-the-job experience and rotation through various parts of the utility. Internal promotion policies would stress career opportunities through internal training and promotion from within the organization.

The personnel policies—low entry level, internal development, and use of employee experience in planning and operations—could not only lead to a pattern of staff stability and high morale, but also to cost-effective operations. Such performance would attract investors to bond offerings of the private utility on U.S., Mexican, or international bond markets.

Low overhead would permit a rate policy that could encourage continual extension of the system to more distant users. Water, wastewater, and drainage rates would be set to encourage underserved rural or colonia areas to joint the system. Subsidies would be drawn from all levels of government to spread the costs and coopt all relevant institutions.

Public participation would be encouraged through existing local institutions. The hydraulic utility would be owned by the communities it serves (and, if appropriate, in part by state and national governments as well), and controlled by representatives of existing local institutions. The participating communities would own shares in the regional utility in proportion to the number of residents.

The regional utilities would not spring full-blown into existence, but be a conscious strategy of the two national governments and the IBWC/CILA. Each national government would have to authorize the operation of each of the binational utilities and each state would have to enact legislation to permit inter-local cooperation with existing state and local agencies. The IBWC/CILA could establish each regional utility through its minutes procedure and provide technical assistance and quality control in lieu of direct regulation from two sets of state and national regulatory agencies.

These regional surface/groundwater supply and wastewater/ drainage utilities could develop the scale, internal expertise, financial stability, and public support necessary for continued improvement of water infrastructure along the border region. The following section describes the basis in precedent and law for establishment and operation of such private regional hydraulic utilities.

Procedures for Creating Regional Hydraulic Utilities

Four steps would be required to create private, binational regional utilities responsible for managing groundwater and surface waters, wastewater treatment and drainage. These are:

1. Each national and state government would authorize through legislation the establishment of each private, binational utility.
2. The IBWC/CILA would approve through its Minute process the creation and operation of the regional utilities, under IBWC/CILA jurisdiction and supervision.
3. The federal, state, and local governments would cooperate in development of separate regional plans for the implementation of each regional hydraulic utility.
4. Citizens in each set of twin cities would vote to integrate all existing water utilities into the new regional binational hydraulic utility.

National and state sovereignty preempts local institutions or private companies from operating as entities across national borders without an explicit grant of authority. For example, it is possible for a streetcar company to operate in the twin metropolitan areas of Ciudad Juárez/El Paso if the U.S. and Mexican national governments separately authorize the company to operate and the state of Texas also authorizes the company. What is required are three separate pieces of legislation by the U.S. Congress, the Mexican Legislature, and the Texas Legislature. Neither national nor state sovereignty would be violated if each unit of government authorizes explicitly the creation of a binational institution.

The jurisdiction of the IBWC/CILA for water resources and water quality issues along the Rio Grande/Río Bravo would not be affected adversely if the commission itself authorizes the creation of each binational hydraulic utility through the minute process created by the Treaty of 1944.[55] A precedent exists for establishing a bilateral water institution not managed by the IBWC/CILA but falling under its jurisdiction: sanitary improvements associated with the Laredo/Nuevo Laredo wastewater collection and treatment project. Minute Order 249 authorizes "the government of Mexico and the state of Tamaulipas to construct and operate and maintain" wastewater infrastructure under the purview of the IBWC/CILA.[56] Explicit authorization by the IBWC/CILA of all powers of each regional utility would also establish the IBWC/CILA as the regulatory authority for quality assurance in lieu of a diverse set of overlapping authorities of federal and/or state agencies in the U.S. and Mexico which might otherwise assert jurisdiction.

Federal, state, and local government could each contribute to regional plans for the development and improvement of water infrastructure within the regions. The regional water utilities in the Netherlands were based around such plans, established even before the utilities began to operate.[57] If regional plans are developed through cooperation of all levels of government and the IBWC/CILA, then there would be a strong presumption as to their long-term rationale and viability.

Diverse private and public water institutions already exist on each side of the twin metropolitan areas. It would be necessary for local voters to integrate those institutions within each regional hydraulic utility. At least within the state of Texas, such a reorganization of authorities is constitutional and provided for through water law.[58] A democratic expression of a majority of each jurisdiction within the binational region would also provide a popular basis for long-term commitment and planning.

Conclusions

Alternate models exist for water and wastewater infrastructure institutions along the border between Mexico and Texas. A legal basis exists for state involvement, local initiative, and even private control of development of hydraulic infrastructure. The choice comes down to three issues: who will pay, plan for, and manage these projects? The current institutional responses through the IBWC/CILA and the EPA/Sedesol have provided too little too late. Although federal agencies within the United States and Mexico have begun to take some initiative through the La Paz agreement, existing environmental plans remain a compendium of current projects of disparate institutions.

The states have begun significant initiatives. However, local citizens will never have the water and wastewater infrastructure they desire unless they decide to create institutions that can plan, manage, and pay for hydraulic improvements. Although federal and state governments may be inclined to subsidize some expenses of improving water infrastructure along the border, regional institutions cannot become effective until they adopt a goal of self-sufficiency using rates that allow financing using competitive capital markets. The creation of institutions that can achieve the infrastructure goals within the limited means of the citizens of the border region promises to be a challenging task during the coming decades.

Notes

1. Treaty Between the United States of America and Mexico, "Utilization of Water of the Colorado and Tijuana rivers and of the Rio Grande," Stat. 1219, TS 994, Article 2, 1944, p. 7.

2. "Agreement Between the United Mexican States and the United States of America on Cooperation for the Protection and Improvement of the Environment in the Border Area," August 1983, Article 12, Para. 2.

3. U.S. Environmental Protection Agency and the Mexican Secretariá de Desarrollo Urbano y Ecología, *Mexican-U.S. Integrated Border Environmental Plan for the Mexico-U.S. Border Area, (First Stage, 1992-1994)*, August 1991 working draft.

4. Three recent papers review a portion of this literature. The papers were each available from the authors prior to publication. Stephen P. Mumme, "Innovation and Reform in Transboundary Resource Management: A Critical Look at the International Boundary and Water Commission, United States and Mexico," presented at the Tri-National Conference on the North American Experience Managing International Transboundary Water Resources, Gasparillo Island, Florida, April 1991 and unpublished paper, Colorado State University, Colorado; Alberto Székely, "Emerging Boundary Environmental Challenges and Institutional Issues: Mexico and the United States," unpublished paper, Colegio de México, Mexico City; and Roberto A. Sánchez, "Public Participation and the International Boundary and Water Commission: Challenges and Options," El Colegio de la Frontera Norte, Tijuana, Mexico.

5. Treaty of Guadalupe Hidalgo of Peace, Friendship, Limits and Settlement between the United States and Mexico, TS 207; 9 Stat. 922; 18 Stat., Part 2, Public Treaties, 1848.

6. International Boundary and Water Commission, *The International Boundary and Water Commission,* United States and Mexico, El Paso, 1989, p. 2.

7. Treaty of 1944, Article 2.

8. Treaty of 1944, Article 2.

9. Treaty of 1944, Article 2.

10. Cruz Ito, U.S. Section of the IBWC, personal communication to Joseph E. Roth, April 17, 1990.

11. Joaquin R. Bustamante, former Commissioner of the Mexican Section of the IBWC, personal communication to Joseph E. Roth, March 30, 1990.

12. Cruz Ito, U.S. Section of the IBWC, personal communication to Joseph E. Roth, March 30, 1990.

13. Treaty of 1944, Article 5.

14. Treaty of 1944, Article 5.

15. Joseph F. Friedkin, former Commissioner of the U.S. Section of the IBWC, personal communication to Joseph E. Roth, March 29, 1990.

16. International Boundary and Water Commission, *Joint Projects of the United States and Mexico ,* El Paso, Texas, 1981, p. 16.

17. Treaty of 1944, Article 7.

18. IBWC, "Joint Projects," page 18.

19. Joseph Friedkin, March 29, 1990.

20. Joseph Friedkin, March 29, 1990.

21. IBWC, "Joint Projects," p. 15.

22. Cruz Ito, March 30, 1990.

23. U.S. Environmental Protection Agency and the Mexican Secretaría de Desarrollo Urbano y Ecología, *Mexican-U.S. Integrated Border Environmental Plan,* 1992-1994, February 1992.

24. EPA and SEDUE, February 1992.

25. Jan Gilbreath Rich, "Planning for the Border's Future: The Mexican-U.S. Integrated Border Environmental Plan," Occasional Paper No. 1, U.S. Mexican Policy Studies Program, LBJ School of Public Affairs, The University of Texas at Austin, February 1992.

26. Treaty of 1944, Articles 2, 6, and 8.

27. Todd Chenoweth, Texas Water Commission, telephone interview with David Hurlbut, July 28, 1992.

28. Jan Gilbreath Rich, *Planning the Border's Future,* p. 21.

29. Todd Chenoweth interview, July 28, 1992.

30. Jan Gilbreath Rich, *Planning the Border's Future,* p. 30. It was not clear as of this writing how much additional federal money would actually be appropriated by Congress for the next fiscal year.

31. R. J. Freeman and M.E. Akers, "Water Rights Adjudication: Introduction and Background," interoffice report, Texas Department of Water Resources, Austin, Texas, February 1979.

32. *Texas Water Code,* Vernon's Texas Water Code Annotated, West Publishing Co., St. Paul, Minnesota, 1988.

33. Norman D. Johns and Richard E. Brown, "Watermaster Administration of Texas State Waters," Presentation to the American Society of Civil Engineers, Water Resources Planning and Management Division Meeting, Fort Worth, Texas, 1990.

34. Andrew Schoolmaster, "Water Marketing in the Lower Rio Grande Valley, Texas," *Professional Geographer,* Vol. 43, No. 3, August 1991, pp. 292-304.

35. Andrew Schoolmaster, "Water Marketing."

36. International Boundary and Water Commission, "Final Environmental Assessment: International Agreement for Improvement of the Quality of the Waters of the Rio Grande at Laredo, Texas/Nuevo Laredo, Tamaulipas," El Paso, Texas, August 1989, p. 1.

37. International Boundary and Water Commission, "Joint Report of the Principal Engineers Concerning Measures that Should be Undertaken to Improve the Quality of the Waters of the Rio Grande at Laredo, Texas/Nuevo Laredo, Tamaulipas," August 1989, p. 2.

38. IBWC, "Final Environmental Assessment," p. 2.

39. IBWC, "Final Environmental Assessment," p. 3.

40. IBWC, "Final Environmental Assessment," p. 4.

41. Fernando Roman, Laredo Water Works Superintendent of Wastewater, personal communication, February 1, 1990.

42. Tomás Rodriguez, Laredo Water Works Assistant Director, personal communication, February 1, 1990.

43. Fernando Roman interview, February 1, 1990.

44. Fernando Roman interview, February 1, 1990.

45. Tomás Rodriguez interview, February 1, 1990.

46. Fernando Roman interview, February 1, 1990.

47. International Boundary and Water Commission, Minute Order No. 279, "Joint Measures to Improve the Quality of the Waters of the Rio Grande at Laredo, Texas/Nuevo Laredo, Tamaulipas," El Paso, Texas, August 28, 1989, p. 4.

48. Telephone conversation between Jim Colby, assistant to U.S. Congressman Kika de la Garza, and Donna Zinke, March 30, 1988.

49. Unpublished news release from the office of Congressman Kika de la Garza, December 16, 1987.

50. Telephone conversation between Jim Colby, assistant to U.S. Congressman Kika de la Garza, and Donna Zinke, March 30, 1988.

51. Rio Grande Pollution Correction Act of 1987, H.R. 2046, 100th Congress, First Session, April 9, 1987.

52. Helen Ingram and David R. White. "The U.S. Section of the International Boundary and Water Commission: Expanding State and Local Involvement, Ambos Nogales and the Need for Change," paper presented at the Tri-National Conference on the North American Experience Managing International Transboundary Water Resources, Gasparilla Island, Florida, April 1991.

53. David J. Eaton, "Extending Rural Water Supplies in Developing Countries," *Journal of the American Water Works Association,* Vol. 77, June 1985, pp. 11, 106.

54. David J. Eaton, "Developing Regional Water Systems in the Netherlands," *Journal of the American Water Works Association,* Vol. 77, June 1985, pp. 70-72.

55. Treaty of 1944, Articles 6 and 8.

56. IBWC, Minute Order No. 249, August 28, 1989, p. 3.

57. David J. Eaton, "Developing Regional Water Systems."

58. Russell K. Hedge, Jr. and David J. Eaton, "Cooperative Service Arrangements," *Southwest and Texas Water Works Journal,* Vol. 67, July 1985, pp. 5-6, 8, and 10.

Chapter 6:
Survey of Attitudes Toward Water Quality

AS MEXICO AND THE UNITED STATES REDUCE BILATERAL trade restrictions and increase joint attention to border environmental issues, one question stands out: how much public support exists for cooperative problem-solving of joint water problems? Federal, state, or local government initiatives cannot be implemented if the people who must live with and finance programs are not supportive. This chapter investigates the knowledge and attitudes of Texas and Mexican border residents toward water quality issues and their willingness to pay for and support solutions to problems.

When considering the wide range of real and potential threats to the environment, water quality often ranks at the top of the public's most serious concerns. A recent survey by The Gallup Poll showed that people worried about water pollution more than other environmental problems.[1] In another survey of three California communities, more than 79 percent of the respondents said they were very concerned about the possibility of contaminated drinking water.[2] Both surveys included water quality as only one of many issues about the environment at risk.

This chapter examines the perceptions and attitudes of people along the Rio Grande/Río Bravo regarding the quality of the water they use. By surveying people in an affected area, one may gauge the broader community knowledge and attitudes which constitute the political dimensions of water quality improvement strategies.

Survey Goals

The extent to which people are informed about an issue such as water quality is important to a policy maker. Support for policies may depend on what the concerned parties know about which water sources are polluted, what is contaminating them, and where information may be obtained. If people are not aware that their water is polluted, they may undervalue projects designed to improve quality. They also might pollute in ways they otherwise would not if they had full information. Thus, the first goal of this survey was to elicit from citizens of one pair of border communities the extent of their knowledge about current water quality conditions.

Closely related to knowledge are respondents' attitudes towards water quality. It is important to know what priority taxpayers would assign water quality compared to other policy issues, and whether they are concerned enough to pay for water improvement projects. Are citizen attitudes about water quality strong or stable enough to affect the policy process? Do United States and Mexican citizens differ in their level of concern? It is also important to look for correlation between knowledge and attitudes. If respondents have never thought about water quality, the attitudes they list may be lukewarm below the surface and would therefore be misleading to the policy process.[3] A community's attitudes must be complemented by full information about trade-offs if policy is to be politically practicable and effective.

Finally, given the residents' knowledge of and attitudes toward water quality problems, the survey sought to identify views of current policy. How do they view the cooperation among agencies and countries? Who do they perceive to be currently involved in the process? Are current policies addressing their concerns? There are many ties between policy preferences, water quality, knowledge, and attitudes.

Sparse financial resources limited the sample of Rio Grande Basin residents to one population center on the border. The sister cities of Laredo and Nuevo Laredo in the border states of Texas and Tamaulipas was selected because of its well-documented water quality problems and the recent Laredo Sanitation and Treatment Project proposal. The Laredo Chamber of Commerce has also worked aggressively in the community on issues of water quality. Because Laredo/Nuevo Laredo has wide exposure to water quality issues, the probability of receiving meaningful responses to survey questions was high.

Methodology

Three different survey methods were considered: mail, telephone, and personal interview. A mail survey is relatively low-cost and can cover a large area. The mail survey method was selected because of time and resource constraints and to contact the broadest and most representative sample possible. The Laredo Water Works and the Comisión Federal de Electricidad in Nuevo Laredo provided complete customer lists from which to draw random samples. The Laredo list included 29,500 households. The Nuevo Laredo list included 10,890 households.

The next step was to determine the sample size needed in order to obtain meaningful results. A sample size of 1,000 households was calculated by using the formula:

$$n = \left\{ \frac{Z\sqrt{pq}}{B} \right\}^2$$

where Z indicates the z-score for the desired level of confidence (in this case, 1.96 for a 95 percent level of confidence); p and q indicate the expected probability of getting a "positive" outcome or a "negative" outcome on an answer (both .5); and B denotes the desired level of accuracy, also called the error bound or error of estimation (±3 percent).

Half of the 1,000 surveys were sent to Laredo, and the other half to Nuevo Laredo. After the surveys were collected, the formula was used again to recalculate the error based on the number of actual responses. This formula was used to solve for B, both in the combined area as well as for each community. Thus the inferences drawn from this survey are accurate with a 95 percent level of confidence. The inferences drawn from Laredo have a ±8 percent degree of accuracy and the ones drawn from Nuevo Laredo have a ±13 percent degree of accuracy.

The survey was designed for simplicity. It had 20 questions, of which fourteen were substantive and six demographic. The survey was set up on four pages, with English and Spanish versions printed back to back. A letter was included in the mailing which described the project purpose and gave instructions for

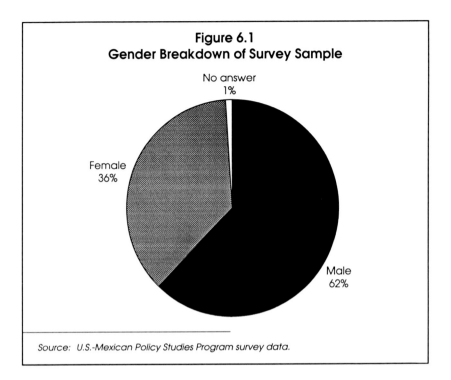

Figure 6.1
Gender Breakdown of Survey Sample

No answer
1%

Female
36%

Male
62%

Source: U.S.-Mexican Policy Studies Program survey data.

returning the survey. Follow-up post cards were sent approximately two weeks after the surveys were mailed.

Results

Survey response was good: 162 surveys (32 percent) were returned from Laredo and 58 surveys (14.2 percent of total delivered) were completed and returned from Nuevo Laredo. A total of 40 percent of the respondents answered the survey in Spanish. Written comments were added by 95 respondents.

Detailed demographic information is shown in Figures 6.1-6.6 and in the appendix to this chapter. Except for education level and income, demographic data varied little between Laredo and Nuevo Laredo. Respondents from Nuevo Laredo tended to have a lower level of education than those from Laredo. Fifty-four percent of Laredo respondents reported a household income below $30,000, compared to 74 percent of Nuevo Laredo respondents. Sixty percent of the respondents were living in

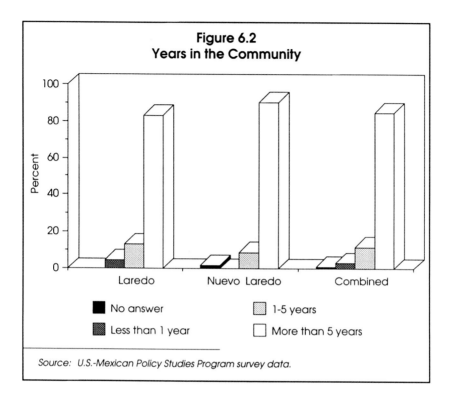

Figure 6.2
Years in the Community

Source: U.S.-Mexican Policy Studies Program survey data.

households with children. A majority of the survey respondents were male. More than 80 percent of the respondents had lived in the community for more than five years.

Respondents' Attitudes about Water Quality

The survey reveals that citizens of the Laredo/Nuevo Laredo community have a high level of concern about water quality: almost 91 percent of the respondents said water quality was a problem. (See Figure 6.7.) Six more questions, intended to ascertain the depth and fragility of this general concern, asked respondents to describe the water they use in their homes, to compare the issue of water quality with other issues, and to comment on their willingness to take action to improve water quality.

Even though all but 5 percent of the total respondents said they got their water from city supplies, 65 percent of the Laredo

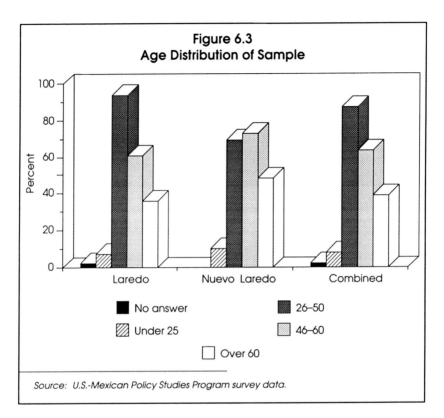

Figure 6.3
Age Distribution of Sample

Source: U.S.-Mexican Policy Studies Program survey data.

respondents and 84 percent of the Nuevo Laredo respondents said they did not think their water was of good quality (Figure 6.8). Several wrote that they purchase bottled water for drinking.

Discoloration was most often cited as the indicator of poor quality (Figure 6.9). Several also noted the presence of particles in the water and a white or cloudy appearance. Twenty-five percent of the respondents (16 percent in Laredo and 50 percent in Nuevo Laredo) said that the water had made them ill. Bad odor was also mentioned by 25 percent of the respondents. Other frequently mentioned reasons were bad taste and too much chlorine or chemicals. Several respondents referred to contamination of the river with sewage, chemicals, or heavy metals. Others cited news reports about increased incidence of cancer in the area.

Respondents were asked to rank the importance of water quality with other issues. Members of the Laredo/Nuevo Laredo

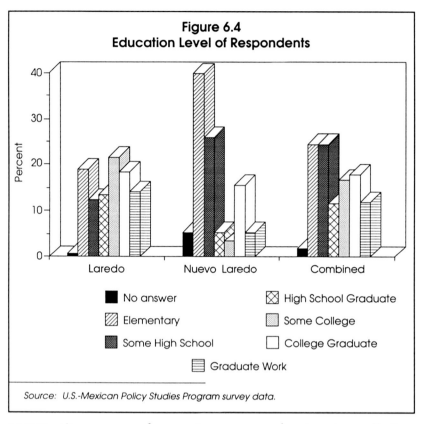

Figure 6.4
Education Level of Respondents

Source: U.S.-Mexican Policy Studies Program survey data.

community expressed a greater concern about water pollution by a margin of three to one. They were then asked to rank a list of issues in order of importance: health care, education, reduction of crime, water quality, economic development, building and repairing roads, and parks. Health care was most often ranked first, followed by education and water quality. In Nuevo Laredo, however, water quality was most frequently cited as the number one concern (Figure 6.10). Combining the marks for first, second, and third rankings, education scored highest followed by health care and water quality—again with water as the most important concern in Nuevo Laredo separately. The next highest issue, reduction of crime, was 14 percentage points behind water quality. Water quality and education received the least number of sixth and seventh place rankings.

Water quality is clearly much on the minds of the citizens in this community. One surprising result of this question was the

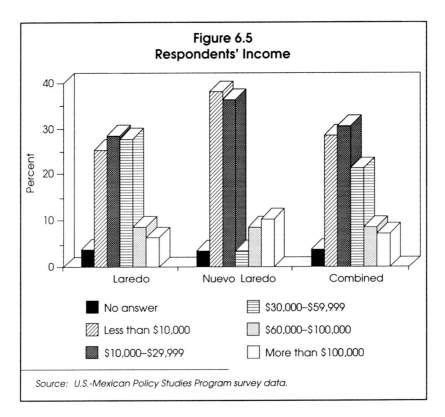

Figure 6.5
Respondents' Income

Legend:
- No answer
- Less than $10,000
- $10,000–$29,999
- $30,000–$59,999
- $60,000–$100,000
- More than $100,000

Source: U.S.-Mexican Policy Studies Program survey data.

relatively low ranking of economic development. Environmental protection often plays a tug-of-war with the desire for community development and economic health, but in this survey only 24 percent of the respondents ranked economic development as one of the top three issues.

Respondents were then asked about taking personal action on the issue. They were asked about willingness to pay for water quality improvements. They were asked specifically about their willingness to pay more for monthly service or to pay higher taxes for water quality improvements. A total of 64 percent of Laredo respondents and 84 percent of Nuevo Laredo respondents (69 percent combined) said that they would indeed be willing to pay more. (See Figure 6.11.) A few respondents qualified their responses. One woman said: "U.S. money for U.S. improvements, not U.S. money for Mexican improvements." A few doubted the government would spend needed funds for water quality. Several respondents commented that they just

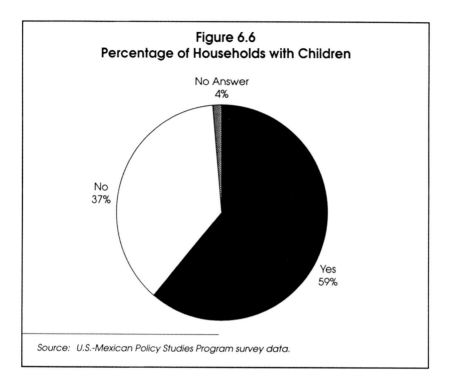

Figure 6.6
Percentage of Households with Children

No Answer
4%

No
37%

Yes
59%

Source: *U.S.-Mexican Policy Studies Program survey data.*

could not afford increased taxes or bills.

The financial position of respondents had no significant effect on a respondent's answer. There is only a .09 correlation between household income and willingness to pay more for water quality. Residents of Nuevo Laredo gave similar responses on the willingness to pay issue. A total of 76 percent said they would pay more. (See Figure 6.12.) It is clear that a significant majority—and more than three-fourths of the Nuevo Laredo respondents—would commit part of their budget to the improvement of their community's water.

Respondents were also asked if they had ever filed a complaint about water quality. Only 14 said they had, and of those, only one was satisfied with the response to the complaint. The low affirmative rate may reflect the fact that residents feel powerless to affect water quality by complaining to a government agency.

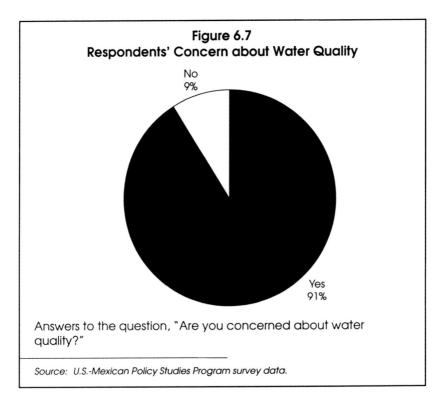

Figure 6.7
Respondents' Concern about Water Quality

No
9%

Yes
91%

Answers to the question, "Are you concerned about water quality?"

Source: U.S.-Mexican Policy Studies Program survey data.

Factors Influencing Attitudes about Water Quality

Many Laredo citizens see the lack of sewage treatment facilities in Nuevo Laredo as a sign that Mexican citizens have a lower level of concern about water quality than do U.S. citizens. Others speculate that the level of concern is related to household income or other demographic factors. But are these explanations sufficient? If a person believes water quality is poor, would demographic characteristics affect the concern? Would government policy reflect citizen interest and concern?

A regression model was used to test the significance of demographic factors or access to accurate information. The dependent variable was an index of concern about water quality derived from weighted responses to five questions. (Table 6.1 shows the questions, answers, and weights.) A respondent answering all questions in the strongest level of concern would score 0.8, while a respondent answering all questions with the weakest level of concern would score 0.

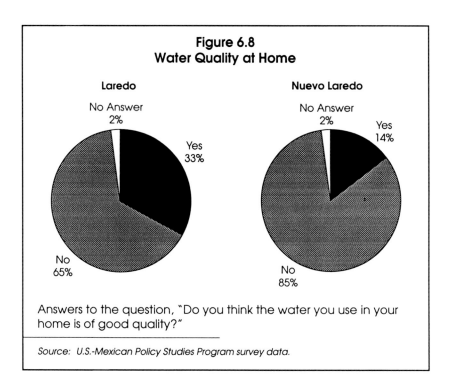

Figure 6.8
Water Quality at Home

Laredo

No Answer 2%

Yes 33%

No 65%

Nuevo Laredo

No Answer 2%

Yes 14%

No 85%

Answers to the question, "Do you think the water you use in your home is of good quality?"

Source: U.S.-Mexican Policy Studies Program survey data.

As a check on the validity of the index, a factor analysis was performed on the five components. The factor score coefficients obtained from the algorithm can indicate how well a set of variables fit together and how well they capture the concept in the dependent variable. If the chosen variables are a cohesive set, they are transformed into z-scores and weighted according to coefficients from the factor analysis. The five components of the index were considered cohesive enough to describe the desired "level of concern" concept. A regression was run with the factor-weighted dependent variable; the results were similar to the regression run with the index dependent variable. The indexed variable was used in the model because it was easier to interpret.

Ten independent variables were chosen to test various demographic attributes, experience, and information access. The variables are listed in Table 6.2. Two demographic variables were expected to be significant: education and children in the household. If people are better educated, they are more likely to have or come into contact with information about water qual-

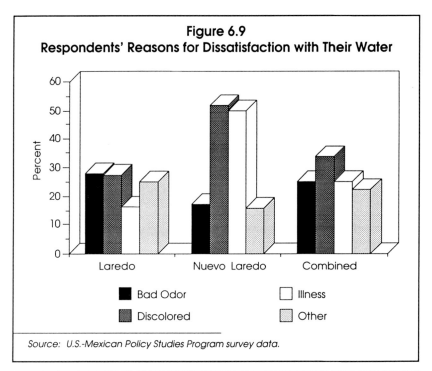

Figure 6.9
Respondents' Reasons for Dissatisfaction with Their Water

Source: U.S.-Mexican Policy Studies Program survey data.

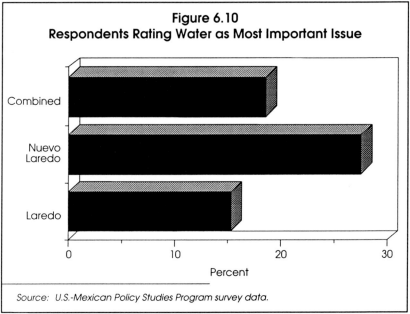

Figure 6.10
Respondents Rating Water as Most Important Issue

Source: U.S.-Mexican Policy Studies Program survey data.

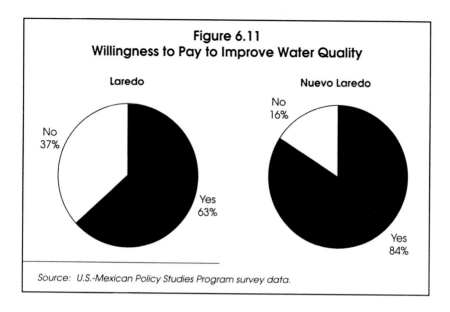

Figure 6.11
Willingness to Pay to Improve Water Quality

Laredo

No 37%

Yes 63%

Nuevo Laredo

No 16%

Yes 84%

Source: U.S.-Mexican Policy Studies Program survey data.

ity, especially if information is not readily available in the community. An added level of concern about water quality may be present in households with children, as parents are concerned for the health and safety of their children. None of the other demographic variables was expected to influence the level of concern in a significant way. A variable was added to assess the influence of sparse information on level of concern (NOINFO). Another term was included to determine whether possession of correct information about water quality was influential (NEEDPLT). If respondents said that a treatment plant was necessary, they were assumed to have some accurate information. A variable representing personal experience with illness or other effects of poor water quality was expected to be significant.

Given the uncertainty of using one piece of information (in this case, NEEDPLT) as a proxy for possession of correct information, two regressions were run, with and without the NEEDPLT variable. The regression results are shown in Table 6.3. The model was not particularly sensitive to the presence of the variable; the magnitudes and signs of all but one parameter estimate remained the same. The Durbin-Watson statistic (1.96) indicated a model free of autocorrelation. Since many of the variables were dummy variables, and since exploratory data analysis had shown concentrated responses, multicollinearity

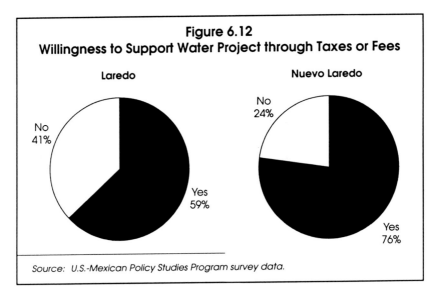

Figure 6.12
Willingness to Support Water Project through Taxes or Fees

Laredo

No 41%

Yes 59%

Nuevo Laredo

No 24%

Yes 76%

Source: U.S.-Mexican Policy Studies Program survey data.

problems could have been expected in the results. A correlation matrix and collinearity diagnostics were run to test for multicollinearity between the explanatory variables, and these showed that the model was free of any significant problems.

Three variables are significant in both models: age (denoted by AGE in Table 6.3), experience with poor water quality (denoted by EXPER), and residency in the United States (denoted by USA). As expected, experience with water quality was a significant factor. Age was a surprisingly significant factor; younger respondents showed more concern. This could indicate that older respondents, probably long-time residents of the community, have seen improvements to the water system over the years. Remembering how bad the water used to be, they perhaps are now less concerned relative to others.

The variable USA was significant and negative; Mexican respondents showed a higher level of concern than U.S. respondents. Even though a majority of all respondents are concerned about water quality, those in Nuevo Laredo were less likely to feel good about the water in their home and were more likely to be willing to pay to improve water quality. This makes sense, however, if personal experience influences the level of concern, as this model indicates. Since Nuevo Laredo residents tend to have more experience with poor water, their concern would be stronger.

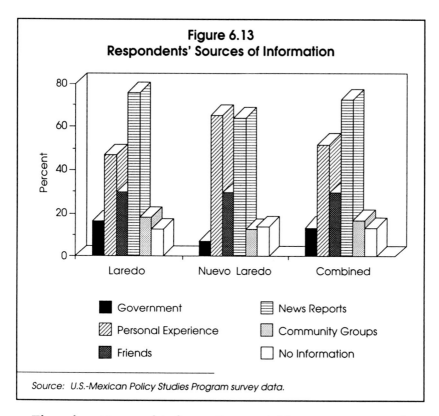

Figure 6.13
Respondents' Sources of Information

Legend:
- Government
- Personal Experience
- Friends
- News Reports
- Community Groups
- No Information

Source: U.S.-Mexican Policy Studies Program survey data.

The education and information variables were expected to have much more influence over level of concern. NOINFO was negative in both models as expected, but was significant at the 90 percent level in the first model and was just shy of that in the second. If the survey is redone, the model would be strengthened by a purer variable to capture lack of information.

There was also some concern about incorporating the "willing to pay" questions in the index because household income might be a significant factor. Then ability to pay would be distorted into willingness to pay. The results of the regression show this is not a problem. All other factors held constant, income does not significantly influence the level of concern, even when concern is measured in part by a willingness to pay higher taxes or utility rates.

The regression results bring out some important points. First, it can be misleading to make assumptions about attitudes towards water quality based on demographic characteristics. The

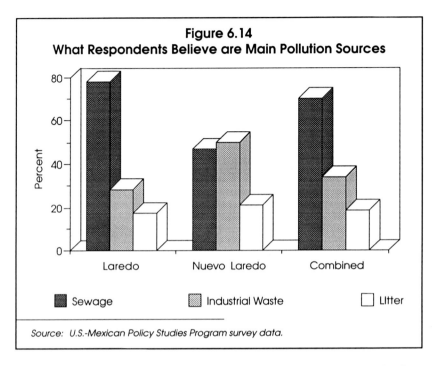

Figure 6.14
What Respondents Believe are Main Pollution Sources

Source: U.S.-Mexican Policy Studies Program survey data.

underlying causes need to be investigated to understand what triggers individual concern. Second, it is fallacious to assume governmental inaction reflects individuals' level of concern. Poor water quality *causes* concern; it does not result from lack of concern.

Respondents' Knowledge about Water Quality

Five questions on the survey were intended to get a sense of the respondent's knowledge and access to information about water quality. The first was designed to discover to what degree respondents felt they had access to information about water quality (Figure 6.13). Respondents were given a list of possible sources of information and were asked to check all that applied. The list included government agencies, personal experience, friends, news reports, and community groups. The final choice was that they didn't receive any information about water quality; 12 percent of the respondents gave this answer. Laredo respondents cited news reports as the most common source of

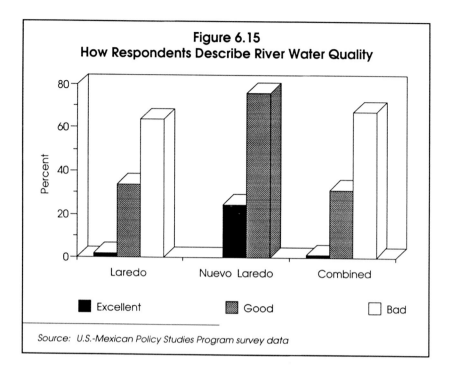

Figure 6.15
How Respondents Describe River Water Quality

Source: U.S.-Mexican Policy Studies Program survey data

information (75.9 percent), followed by personal experience, friends, and community groups. Nuevo Laredo respondents gave a similar ranking, except that personal experience edged out news as the most common information source.

The other questions in this category were an attempt to validate the accuracy of respondents' information about water quality in this section of the Rio Grande/Río Bravo. Respondents were asked to identify the major source of pollution in the river; 78 percent of the Laredo respondents correctly answered sewage (see Figure 6.14). However, 50 percent of the Nuevo Laredo respondents listed industrial waste as the major source—perhaps due to sensitivity about the maquiladora plants. Sixty-seven percent of the respondents said the quality of the water in this part of the river was bad (see Figure 6.15).

Two questions in the survey dealt with knowledge about a current proposal to build a wastewater treatment plant in Nuevo Laredo. These were placed in the survey to ascertain whether respondents had information about specific problems and proposed solutions. A total of 72 percent of the respondents said

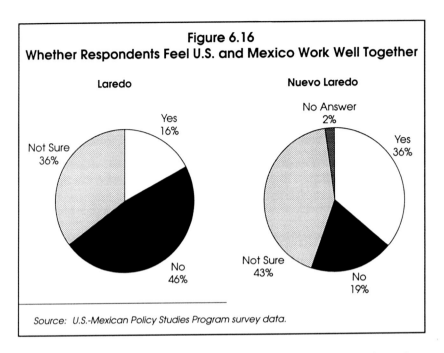

Figure 6.16
Whether Respondents Feel U.S. and Mexico Work Well Together

Laredo

Nuevo Laredo

Source: U.S.-Mexican Policy Studies Program survey data.

they were aware of the proposal, and 97 percent said such a project was necessary.

The survey results show that citizens in the community of Laredo/Nuevo Laredo are for the most part getting accurate information about the state of the river. Yet more than 12 percent of the respondents reported that they did not receive information about water quality, which implies that public officials still need to communicate with a significant part of the population.

Respondents' Attitudes about Current Water Quality Policy

The last goal of the survey was to learn what people believe about current water quality policy and the effectiveness of policy-making bodies. The results indicate the respondents were not seeing enough visible signs of effort in the area of water quality management.

Respondents were asked to rate the attention given by the government to water quality. Overall, 59 percent (and 79.3 percent of the Nuevo Laredo respondents) felt that government was

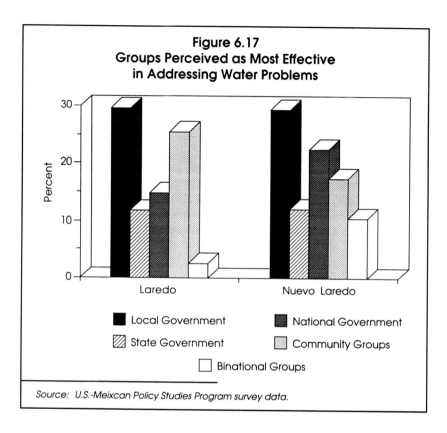

Figure 6.17
**Groups Perceived as Most Effective
in Addressing Water Problems**

Source: U.S.-Meixcan Policy Studies Program survey data.

not giving the subject enough attention. Respondents were also skeptical about whether the United States and Mexico work well in solving water problems. Thirty-eight percent felt that they did not work well together and another 38 percent were not sure (see Figure 6.16). Finally, respondents were asked to rank in order of their effectiveness local government, state government, national government, community groups, and binational groups. Local government was perceived as the most effective. Laredo respondents also ranked community groups high (see Figure 6.17). This is consistent with the strong leadership of the Chamber of Commerce and other groups in working towards water quality solutions.

Conclusion

Citizens of the Laredo/Nuevo Laredo community are clearly concerned about water quality. They are informed about the topic, they have experienced personally the effect of poor water quality, and they are prepared to help in solving the problem. In fact, they are willing to pay to help correct the problem, especially if they believe that government would pursue effective action. Public officials and community leaders have a challenge to respond to the level of concern in the community, but so far residents' expectations for government action are not being met. At the same time, however, these expectations give community leaders an enormous political opportunity to make progress in improving water quality in Laredo/Nuevo Laredo. There appears to be a deep and widespread support on both sides of the border for action to improve the quality of the Rio Grande/Río Bravo.

Table 6.1
Index of Concern about Water Quality

Question	Answer	Score
1. Are you concerned about water quality in your community?	yes	1.00
	no	0.00
5. Would you be willing to pay? (general)	yes	0.50
	no	0.00
8. Have you ever filed a complaint?	yes	1.00
	no	0.00
10. Rank issues in importance to you.	water 1st	1.00
	2nd or 3rd	0.75
	4th or 5th	0.50
	6th or 7th	0.25
14. Would you be willing to pay? (specific)	yes	0.50
	no	0.00

Source: U.S.-Mexican Policy Studies Program survey data.

Table 6.2
Description and Expected Sign of the Dependent Variables

	Expected Significance	Expected Sign
LENGTH Length of residency in the community.	no	+
MALE Dummy variable equal to 1 when respondent is male.	no	?
EDUC Mean value of the education category selected by the respondent.	yes	+
AGE Mean value of the age category selected by the respondent.	no	+
INCOME Mean value of the household income category selected by the respondent.	no	+
CHILD Dummy variable equal to 1 when respondent lives in a household with children.	yes	+
NOINFO Dummy variable equal to one if respondent indicated they did not receive information about water quality.	yes	-
EXPER Dummy variable equal to one if respondent indicated a reason for dissatisfaction with the water quality in their home.	yes	+
USA Dummy variable equal to one if the respondent is a U.S. resident.	no	?
NEEDPLT Dummy variable equal to one if the respondent indicated that a treatment plant was needed in Nuevo Laredo.	yes	+

Source: U.S.-Mexican Policy Studies Program survey data.

Table 6.3
Regression Results

Variable	Model 1 Coefficient	Model 1 T-score	Model 2 Coefficient	Model 2 T-score
LENGTH	.00716	.047	.0007	.197
MALE	.0034	.18	-.0011	-.060
EDUC	.0043	.648	.0054	.813
AGE	-.002	-2.69*	-.0021	-2.428*
SALARY	-1.079e-07	-.309	-1.565e-07	-.448
CHILD	-.012	-.595	-.01635	-.801
NOINFO	-.057	-1.9**	-.049	-1.603
NEEDPLT	(variable excluded from model)		.091	1.274
EXPER	.084	4.063*	.080	3.792*
USA	-.073	-3.29*	-.077	-3.394*
Adjusted R²	.1987		.1757	
191 observations				

* Significant at the 95 percent confidence level
** Significant at the 90 percent confidence level

Source: U.S.-Mexican Policy Studies Program survey data.

Notes

1. Andrew Kohut and James Shriver, "The Environment: Environment Regaining a Foothold on the National Agenda," *Gallop Report*, No. 285, June 1989, pp. 2-12.

2. Mark Pilisuk, "Public Perception of Technological Risk," *Social Science Journal*, Vol. 23, No. 4, 1987, pp. 403-13.

3. Earle R. Babbie, *Survey Research Methods*, Belmont, California: Wadsworth Publishing Co. Inc., 1973, p. 143.

Appendix:
Survey Data

THIS APPENDIX CONTAINS THE RAW RESPONSE DATA FOR each question on the survey referred to in Chapter Six. Percentages are shown in parentheses, and refer to all responses from Laredo (first column), all responses from Nuevo Laredo (second column), and the combined total for both cities (third column). Due to rounding, percentages may not add up to 100.

Surveys mailed	1,000
Undelivered surveys	95
Surveys returned	220
Surveys from Laredo, Texas	162
Surveys from Nuevo Laredo, Tamaulipas	58
Surveys not returned	685

	Laredo, Texas		Nuevo Laredo, Tamaulipas		Total	
Answered in English	130	(80.2%)	2	(3.4%)	132	(60.0%)
Answered in Spanish	32	(19.8%)	56	(96.6%)	88	(40.0%)

1. Are you concerned about water quality in your community?

	Laredo, Texas		Nuevo Laredo, Tamaulipas		Total	
No answer	1	(0.6%)	2	(3.4%)	3	(1.4%)
No	14	(8.6%)	3	(5.2%)	17	(7.7%)
Yes	147	(90.7%)	53	(91.4%)	200	(90.9%)

2. Do you think that the water you use in your home is of good quality?

	Laredo, Texas		Nuevo Laredo, Tamaulipas		Total	
No answer	3	(1.9%)	1	(1.7%)	4	(1.8%)
No	106	(65.4%)	49	(84.5%)	155	(70.5%)
Yes	53	(32.7%)	8	(13.8%)	61	(27.7%)

	Laredo	**Nuevo Laredo**	**Total**

If you answered no, is this because

	Laredo	Nuevo Laredo	Total
Odor	45 (27.8%)	10 (17.2%)	55 (25.0%)
Discolored	44 (27.2%)	30 (51.7%)	74 (33.6%)
Illness	26 (16.0%)	29 (50.0%)	55 (25.0%)
Other	40 (24.7%)	9 (15.5%)	49 (22.3%)

3. How would you describe the quality of the water in this part of the Rio Bravo/Rio Grande River?

	Laredo	Nuevo Laredo	Total
Excellent	2 (1.2%)	0 (0.0%)	2 (0.9%)
Good	54 (33.3%)	14 (24.1%)	68 (30.9%)
Bad	103 (63.6%)	44 (75.9%)	147 (66.8%)

4. What do you think is the major source of pollution in the river?

	Laredo	Nuevo Laredo	Total
Sewage	127 (78.4%)	27 (46.6%)	154 (70.0%)
Industrial waste	46 (28.4%)	29 (50.0%)	75 (34.1%)
Litter	28 (17.3%)	12 (20.7%)	40 (18.2%)

5. Would you be willing to pay more for monthly service or pay higher taxes if that money were used to improve water quality?

	Laredo	Nuevo Laredo	Total
No answer	3 (1.9%)	1 (1.7%)	4 (1.8%)
No	56 (34.6%)	8 (13.8%)	64 (29.1%)
Yes	103 (63.6%)	49 (84.5%)	152 (69.1%)

6. What is more of a concern to you?

	Laredo	Nuevo Laredo	Total
Water pollution	122 (75.3%)	41 (70.7%)	163 (74.1%)
Water scarcity	39 (24.1%)	20 (34.5%)	59 (26.8%)
Neither issue	4 (2.5%)	0 (0.0%)	4 (1.8%)

7. Where do you get your information about water quality in your community?

	Laredo	Nuevo Laredo	Total
Government agency	26 (16.0%)	4 (6.9%)	30 (13.6%)
Personal experience	76 (46.9%)	38 (65.5%)	114 (51.8%)
Friends	48 (29.6%)	17 (29.3%)	65 (29.5%)
News reports	123 (75.9%)	37 (63.8%)	160 (72.7%)

	Laredo	Nuevo Laredo	Total
Community groups	29 (17.9%)	7 (12.1%)	36 (16.4%)
No information	20 (12.3%)	8 (13.8%)	28 (12.7%)

8. Have you ever filed a complaint about water quality?

No answer	1 (0.6%)	1 (1.7%)	2 (0.9%)
No	154 (95.1%)	49 (84.5%)	203 (92.3%)
Yes	7 (4.3%)	8 (13.8%)	15 (6.8%)

If yes, were you satisfied with the response?

No	6 (85.7%)	8 (100%)	14 (93.3%)
Yes	1 (14.3%)	0 (0.0%)	1 (6.7%)

9. Do you get your water from the city water system?

No answer	2 (1.2%)	2 (3.4%)	4 (1.8%)
No	2 (1.2%)	3 (5.2%)	5 (2.3%)
Yes	158 (97.5%)	53 (91.4%)	211 (95.5%)

10. How important are these issues to you? Please rank them from 1 to 7, with 1 being the most important issue to you.

Health care

No answer	14 (8.6%)	5 (8.6%)	19 (8.6%)
1	58 (35.8%)	14 (24.1%)	72 (32.7%)
2	32 (19.8%)	18 (31.0%)	50 (22.7%)
3.	18 (11.1%)	8 (13.8%)	26 (11.8%)
4	21 (13.0%)	6 (10.3%)	27 (12.3%)
5	12 (7.4%)	3 (5.2%)	15 (6.8%)
6	3 (1.9%)	3 (5.2%)	6 (2.7%)
7	4 (2.5%)	1 (1.7%)	5 (2.3%)

Education

No answer	17 (10.5%)	6 (10.3%)	23 (10.5%)
1	46 (28.4%)	12 (20.7%)	58 (26.4%)
2	39 (24.1%)	7 (12.1%)	46 (20.9%)
3	39 (24.1%)	11 (19.0%)	50 (22.7%)
4	14 (8.6%)	16 (27.6%)	30 (13.6%)
5	4 (2.5%)	5 (8.6%)	9 (4.1%)
6	1 (0.6%)	1 (1.7%)	2 (0.9%)
7	2 (1.2%)	0 (0.0%)	2 (0.9%)

	Laredo		Nuevo Laredo		Total	
Reduction of crime						
No answer	16	(9.9%)	5	(8.6%)	21	(9.5%)
1	20	(12.3%)	5	(8.6%)	25	(11.4%)
2	25	(15.4%)	6	(10.3%)	31	(14.1%)
3	31	(19.1%)	17	(29.3%)	48	(21.8%)
4	39	(24.1%)	7	(12.1%)	46	(20.9%)
5	22	(13.6%)	9	(15.5%)	31	(14.1%)
6	5	(3.1%)	3	(5.2%)	8	(3.6%)
7	4	(2.5%)	6	(10.3%)	10	(4.5%)
Water quality						
No answer	14	(8.6%)	6	(10.3%)	20	(9.1%)
1	25	(15.4%)	16	(27.6%)	41	(18.6%)
2	32	(19.8%)	13	(22.4%)	45	(20.5%)
3	38	(23.5%)	11	(19.0%)	49	(22.3%)
4	32	(19.8%)	7	(12.1%)	39	(17.7%)
5	17	(10.5%)	4	(6.9%)	21	(9.5%)
6	3	(1.9%)	1	(1.7%)	4	(1.8%)
7	1	(0.6%)	0	(0.0%)	1	(0.5%)
Economic development						
No answer	6	(9.9%)	6	(10.3%)	22	(10.0%)
1	12	(7.4%)	5	(8.6%)	17	(7.7%)
2	15	(9.3%)	5	(8.6%)	20	(9.1%)
3	12	(7.4%)	4	(6.9%)	16	(7.3%)
4	27	(16.7%)	12	(20.7%)	39	(17.7%)
5	55	(34.0%)	23	(39.7%)	78	(35.5%)
6	17	(10.5%)	2	(3.4%)	19	(8.6%)
7	8	(4.9%)	1	(1.7%)	9	(4.1%)
Building and repairing roads						
No answer	17	(10.5%)	6	(10.3%)	23	(10.5%)
1	3	(1.9%)	3	(5.2%)	6	(2.7%)
2	3	(1.9%)	3	(5.2%)	6	(2.7%)
3	4	(2.5%)	0	(0.0%)	4	(1.8%)
4	7	(4.3%)	4	(6.9%)	11	(5.0%)
5	20	(12.3%)	6	(10.3%)	26	(11.8%)
6	95	(58.6%)	32	(55.2%)	127	(57.7%)
7	13	(8.0%)	4	(6.9%)	17	(7.7%)

	Laredo		Nuevo Laredo		Total	
Parks						
No answer	17	(10.5%)	6	(10.3%)	23	(10.5%)
1	1	(0.6%)	0	(0.0%)	1	(0.5%)
2	0	(0.0%)	1	(1.7%)	1	(0.5%)
3	2	(1.2%)	0	(0.0%)	2	(0.9%)
4	3	(1.9%)	0	(0.0%)	3	(1.4%)
5	11	(6.8%)	2	(3.4%)	13	(5.9%)
6	17	(10.5%)	9	(15.5%)	26	(11.8%)
7	111	(68.5%)	40	(69.0%)	151	(68.6%)

11. How much attention is the government giving to water quality?

	Laredo		Nuevo Laredo		Total	
No answer	7	(4.3%)	0	(0.0%)	7	(3.2%)
1	2	(1.2%)	2	(3.4%)	4	(1.8%)
2	1	(0.6%)	0	(0.0%)	1	(0.5%)
3	37	(22.8%)	8	(13.8%)	45	(20.5%)
4	30	(18.5%)	2	(3.4%)	32	(14.5%)
5	85	(52.5%)	46	(79.3%)	131	(59.5%)

12. Do you think Mexico and the United States work well together to solve water problems?

	Laredo		Nuevo Laredo		Total	
No answer	0	(0.0%)	1	(1.7%)	1	(0.5%)
No	74	(45.7%)	11	(19.0%)	85	(38.6%)
Yes	29	(17.9%)	21	(36.2%)	50	(22.7%)
Not sure	59	(36.4%)	25	(43.1%)	84	(38.2%)

13. How effective are these groups at addressing water problems? Please rank them from 1 to 5, with 1 being the most effective group.

	Laredo		Nuevo Laredo		Total	
Local government						
No answer	27	(16.7%)	9	(15.5%)	36	(16.4%)
1	48	(29.6%)	17	(29.3%)	65	(29.5%)
2	40	(24.7%)	13	(22.4%)	53	(24.1%)
3	24	(14.8%)	9	(15.5%)	33	(15.0%)
4	11	(6.8%)	3	(5.2%)	14	(6.4%)
5	12	(7.4%)	7	(12.1%)	19	(8.6%)

	Laredo	**Nuevo Laredo**	**Total**
State government			
No answer	35 (21.6%)	8 (13.8%)	43 (19.5%)
1	19 (11.7%)	7 (12.1%)	26 (11.8%)
2	40 (24.7%)	17 (29.3%)	57 (25.9%)
3	41 (25.3%)	15 (25.9%)	56 (25.5%)
4	24 (14.8%)	8 (13.8%)	32 (14.5%)
5	3 (1.9%)	3 (5.2%)	6 (2.7%)
National government			
No answer	33 (20.4%)	10 (17.2%)	43 (19.5%)
1	24 (14.8%)	13 (22.4%)	37 (16.8%)
2	18 (11.1%)	4 (6.9%)	22 (10.0%)
3	28 (17.3%)	17 (29.3%)	45 (20.5%)
4	33 (20.4%)	8 (13.8%)	41 (18.6%)
5	26 (16.0%)	6 (10.3%)	32 (14.5%)
Community groups			
No answer	32 (19.8%)	13 (22.4%)	45 (20.5%)
1	41 (25.3%)	10 (17.2%)	51 (23.2%)
2	14 (8.6%)	4 (6.9%)	18 (8.2%)
3	23 (14.2%)	2 (3.4%)	25 (11.4%)
4	36 (22.2%)	10 (17.2%)	46 (20.9%)
5	16 (9.9%)	19 (32.8%)	35 (15.9%)
Binational groups			
No answer	35 (21.6%)	14 (24.1%)	49 (22.3%)
1	4 (2.5%)	6 (10.3%)	10 (4.5%)
2	11 (6.8%)	8 (13.8%)	19 (8.6%)
3	11 (6.8%)	1 (1.7%)	12 (5.5%)
4	22 (13.6%)	14 (24.1%)	36 (16.4%)
5	79 (48.8%)	15 (25.9%)	94 (42.7%)

14. Are you aware of a proposal to build a wastewater treatment plant in Nuevo Laredo?

	Laredo	Nuevo Laredo	Total
No	43 (26.5%)	19 (32.8%)	62 (28.2%)
Yes	119 (73.5%)	39 (67.2%)	158 (71.8%)

	Laredo	Nuevo Laredo	Total

Do you consider such project necessary?

	Laredo		Nuevo Laredo		Total	
No answer	2	(1.2%)	0	(0.0%)	2	(0.9%)
No	1	(0.6%)	2	(3.4%)	3	(1.4%)
Yes	159	(98.1%)	56	(96.6%)	215	(97.7%)

Would you be willing to support such project through taxes or fees?

	Laredo		Nuevo Laredo		Total	
No answer	4	(2.5%)	3	(5.2%)	7	(3.2%)
No	62	(38.3%)	11	(19.0%)	73	(33.2%)
Yes	96	(59.3%)	44	(75.9%)	140	(63.6%)

15. How long have you lived in the community?

	Laredo		Nuevo Laredo		Total	
No answer	0	(0.0%)	1	(1.7%)	1	(0.5%)
Less than one year	7	(4.3%)	0	(0.0%)	7	(3.2%)
1–5 years	21	(13.0%)	5	(8.6%)	26	(11.8%)
Over 5 years	134	(82.7%)	52	(89.7%)	186	(84.5%)

16. Are you

	Laredo		Nuevo Laredo		Total	
No answer	0	(0.0%)	3	(5.2%)	3	(1.4%)
Male	101	(62.3%)	36	(62.1%)	137	(62.3%)
Female	61	(37.7%)	19	(32.8%)	80	(36.4%)

17. What is the last level of school you completed?

	Laredo		Nuevo Laredo		Total	
No answer	1	(0.6%)	3	(5.2%)	4	(1.8%)
1st level:	31	(19.1%)	23	(39.7%)	54	(24.5%)
2nd level:	20	(12.3%)	15	(25.9%)	54	(24.5%)
3rd level:	22	(13.6%)	3	(5.2%)	25	(11.4%)
4th level:	35	(21.6%)	2	(3.4%)	37	(16.8%)
5th level:	30	(18.5%)	9	(15.5%)	39	(17.7%)
6th level:	23	(14.2%)	3	(5.2%)	26	(11.8%)

18. In which of these groups do you belong?

	Laredo		Nuevo Laredo		Total	
No answer	2	(1.2%)	0	(0.0%)	2	(0.9%)
Under 25	6	(3.7%)	3	(5.2%)	9	(4.1%)
26–45	76	(46.9%)	20	(34.5%)	96	(43.6%)
46–60	49	(30.2%)	21	(36.2%)	70	(31.8%)
Over 60	29	(17.9%)	14	(24.1%)	43	(19.5%)

	Laredo	**Nuevo Laredo**	**Total**

19. Which of the following comes closest to your total family income last year?

	Laredo	Nuevo Laredo	Total
No answer	6 (3.7%)	2 (3.4%)	8 (3.6%)
Under $10,000	41 (25.3%)	22 (37.9%)	63 (28.6%)
$10,000–$29,999	46 (28.4%)	21 (36.2%)	67 (30.5%)
$30,000–$59,999	35 (27.8%)	2 (3.4%)	47 (21.4%)
$60,000–$100,000	14 (8.6%)	5 (8.6%)	19 (8.6%)
Over $100,000	10 (6.2%)	6 (10.3%)	16 (7.3%)

20. Are there children living in your household?

	Laredo	Nuevo Laredo	Total
No answer	4 (2.5%)	5 (8.6%)	9 (4.1%)
No	60 (37.0%)	21 (36.2%)	81 (36.8%)
Yes	98 (60.5%)	32 (55.2%)	130 (59.1%)